Studies in Computational Intelligence

Volume 494

Series Editor

J. Kacprzyk, Warsaw, Poland

For further volumes:
http://www.springer.com/series/7092

Stefania Montani · Lakhmi C. Jain
Editors

Successful Case-based Reasoning Applications-2

 Springer

Editors
Stefania Montani
DISIT, Computer Science Institute
Universita' del Piemonte Orientale
Alessandria
Italy

Lakhmi C. Jain
Faculty of Education, Science, Technology
   & Mathematics
University of Canberra
Canberra
Australia

ISSN 1860-949X   ISSN 1860-9503   (electronic)
ISBN 978-3-662-52136-6   ISBN 978-3-642-38736-4   (eBook)
DOI 10.1007/978-3-642-38736-4
Springer Heidelberg New York Dordrecht London

Preface

Case-based reasoning (CBR) is an Artificial Intelligence (AI) technique that combines problem solving with continuous learning in advanced decision support systems. CBR collects and exploits the experiential knowledge embedded in previously encountered and solved situations, which are known as *cases*. Thus, by using CBR, the difficulties of knowledge acquisition and of knowledge representation are often lessened. CBR research is now mature and very active, as testified by the large number of theoretical and applicative works that can be found in the literature.

In this book, we have collected a selection of papers on very recent CBR applications. Very interestingly, most of them can be classified as Process-oriented Case-based Reasoning (PO-CBR) works, since they present applications in the fields of business process management, software process reuse, and trace-based reasoning. Such a convergence testifies the growing interest of the CBR community towards this specific research direction. Indeed, also the non-PO-CBR works in this book have some relations with the topic, specifically as regards case acquisition and case representation.

Some of these contributions propose methodological solutions that, even though applied to a specific domain, could be properly adapted to different applications as well. Other chapters report on frameworks and tools that have now reached a very mature development stage.

The chapters thus testify the flexibility of CBR, its capability to solve or handle issues which would be too difficult to manage with other classical AI methods and techniques, and, in conclusion, its applicability in fields where experiential knowledge plays a key role in decision making.

We believe that this book will be valuable to the application engineers, scientists, professor and students who wish to use case-based reasoning paradigms.

We wish to express our gratitude to the authors and reviewers for their contribution.

Alessandria, Italy   Stefania Montani
Canberra, Australia   Lakhmi C. Jain

Contents

Chapter 1
Case-Based Reasoning Systems

Stefania Montani and Lakhmi C. Jain

Abstract This book reports on a set of recently implemented intelligent systems, having the case-based reasoning (CBR) methodology as their core. The selected works witness the heterogeneity of the domains in which CBR can be exploited, but also reveal some common directions that are clearly emerging in this specific research area. The present chapter provides a brief introduction to CBR, for readers unfamiliar with the topic. It then summarizes the main research contributions that will be presented in depth in the following chapters of this book.

1 Introduction

Case-based reasoning (CBR) [1, 10] is an Artificial Intelligence (AI) technique meant to provide automatic reasoning capabilities, while allowing continuous learning, in advanced decision support systems. Specifically, CBR exploits the experiential knowledge collected on previously encountered and solved situations, which are known as *cases*.

The reasoning process can be summarized using the following four basic steps. These are known as the *CBR cycle* (Fig. 1), or as the four *"res"* [1]. The procedure is to:

(1) *retrieve* the most similar case(s), with respect to the current input situation, contained in the case repository, which is known as the *case base*;

(2) *reuse* them, or more precisely their solutions, in order to solve the new problem; some adaptation may be required at this stage;

S. Montani
DISIT, Computer Science Institute, Universita' del Piemonte Orientale,
Viale Michel 11, 15121 Alessandria, Italy
e-mail: stefania.montani@unipmn.it

L. C. Jain (✉)
Faculty of Education, Science, Technology & Mathematics, University of Canberra,
Canberra, ACT 2601, Australia
e-mail: Lakhmi.Jain@unisa.edu.au

S. Montani and L. C. Jain (eds.), *Successful Case-based Reasoning Applications-2*,
Studies in Computational Intelligence 494, DOI: 10.1007/978-3-642-38736-4_1,
© Springer-Verlag Berlin Heidelberg 2014

Fig. 1 The case-based
reasoning cycle

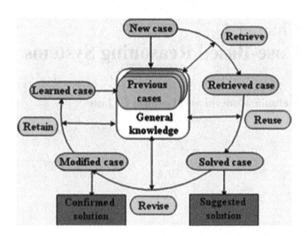

(3) *revise* the proposed new solution (if necessary), to build a new case;

(4) *retain* the new case for possible future problem solving.

In many application domains it is also common to find CBR tools which are able to extract relevant knowledge, but leave the user the responsibility of providing an interpretation and of producing the final decision: steps *reuse* and *revise* are not implemented. In fact even *retrieval* alone may be able to significantly support the reasoning task [16].

As previously stated, CBR not only supports reasoning, but combines problem solving with continuous learning. Indeed, CBR relies on experiential knowledge, in the form of past problem/solution patterns. It does not aim to generalize from past examples, and to learn general rules/models from them, as it happens in other AI reasoning techniques. Indeed, CBR keeps and exploits the specific instances of problems which have been collected in the past (almost) "as they are". By using CBR, the difficulties of knowledge acquisition are therefore often lessened, since the system can automatically and progressively learn new knowledge by means of the *retain* step. The case library then grows in time, and more and more representative examples can be retrieved. This makes it easier to find an appropriate solution to the present problem using this paradigm. This however also implies that proper case base maintenance policies have to be defined.

The first publications on CBR appeared in the 80s. It is therefore a relatively "young" methodology. Nevertheless, CBR research has soon become very active, as testified by the large number of theoretical and applicative works.

Theoretical works have investigated all steps of the CBR cycle [6, 8], with particular emphasis on the design of similarity measures and of fast retrieval strategies by the use of memory organization [5, 17], on adaptation and revision techniques [14], and on case base maintenance [11].

Applications have been proposed for use in several domains [3, 4]. These include legal reasoning [7], health sciences [9, 13], and recommender systems [2], citing only a few examples.

Additionally, in recent years, some new topics are becoming strong areas of interest in the CBR community, i.e. process-oriented reasoning, provenance and traces. These topics have close relationships. Process-oriented reasoning focuses largely on workflows, which define sequences of actions for use and reuse. Provenance captures the results of such sequences, providing a resource for capturing workflow cases and for analyzing how action sequences may be refined. Traces capture the results of action sequences generated "on the fly," and must contend with how to generate useful cases from execution information. CBR applications in the fields of business process management, software process reuse, and trace-based reasoning follow under this umbrella. We will refer to this research direction as Process-oriented Case-based Reasoning (PO-CBR) [12].

Very interestingly, several papers among the ones we selected for publication in this book, belong to PO-CBR. Details of the collected contributions are given in the next section.

2 Chapters Included in the Book

In the following chapters, we present a selection of six very interesting CBR approaches, four of which can be clearly classified as works in the area of PO-CBR; interestingly, the other two have some relations with this domain as well.

The first two contributions are more methodology-oriented, while the others describe CBR tools and frameworks, some of which have already reached a significant development stage, since the authors have been working on their implementation for several years.

The first three chapters deploy CBR for business process management applications.

Chapter 2 (A study of two-phase retrieval for Process-Oriented Case-Based Reasoning, by J. Kendall Morwick and D. Leake) analyses the advantages and issues of exploiting two-phase retrieval methods, which first identify candidate cases with a rather inexpensive retrieval phase, and then apply a more expensive strategy to rank the selected cases. In the paper, these methods are specifically applied to PO-CBR case bases, where cases are structured as graphs. The paper examines the performance of each phase, and provides general lessons for how to design and deploy two-phased retrieval systems in this domain.

Chapter 3 (Non-exhaustive trace retrieval for managing stroke patients, by S. Montani and G. Leonardi) affords the problem of optimizing retrieval performances in PO-CBR too. However, in this contribution, the focus is not on structured cases, but on traces, where only sequence is represented as a control flow relation among actions. Nevertheless, the presence of temporal information in traces makes distance definition non trivial. A proper metric, and different non exhaustive retrieval techniques, are presented in the paper.

Chapter 4 (Evaluating a Case-based reasoning architecture for the intelligent monitoring of business workflows, by S. Kapetanakis and M. Petridis) presents a

CBR tool for intelligent monitoring of business workflows. The framework stores and retrieves a set of business workflow execution traces, by applying proper indexing techniques. This enables the reuse of associated knowledge about the workflow execution into the new target case.

Chapter 5 (The COLIBRI platform: tools, features and working examples, by J.A. Recio-Garcia, B. Diaz-Agudo and P.A. Gonzalez-Calero) presents the Colibri framework, summarizing its main features and illustrating the generic use cases it can deal with. Colibri is an open source framework that allows to build CBR system prototypes. It provides a readily usable implementation of common techniques in CBR, that can be reused and extended properly. The contribution thus falls in the area of PO-CBR too, and specifically in software process reuse support.

The last two chapters are not PO-CBR works. Indeed, they are applications of CBR to very different domains, i.e., ancient manuscript annotation, and cooking. However, some PO-CBR techniques are described and exploited in these contributions as well, namely trace-based reasoning, and workflow cases retrieval.

Specifically, **Chap.** 6 (Case-based reasoning to support annotating manuscripts in digital archives, by R. Doumat), presents a tool for supporting annotation of ancient manuscripts. A CBR approach allows to retrieve past traces, recording annotation sessions. Traces are structured in reusable sections called episodes. The tool recommends proper actions to be undertaken in the current episode, on the basis of the

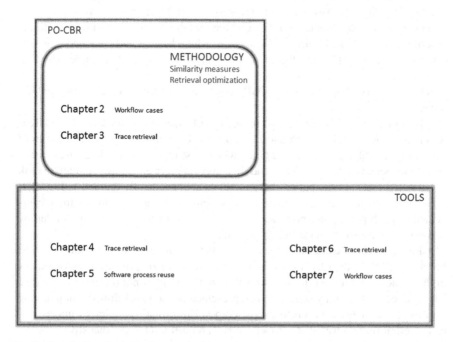

Fig. 2 Contributions in the book, organized by topic

retrieved traces content. This allows to accelerate the annotation process and correct user mistakes.

Chapter 7 (Taaable: a Case-based system for personalized cooking, by A. Cordier, J. Lieber, E. Nauer, F. Badra, J. Cojan, V. Dufour-Lussier, E. Gaillard, L. Infante-Blanco, P. Molli, A. Napoli, H. Skaf-Molli) summarizes the main features of a CBR system that uses a recipe book as a case base to answer cooking queries. The system combines various methods from knowledge-based systems: CBR, knowledge representation, knowledge acquisition and discovery, knowledge management, and natural language processing. In particular, the tool exploits a method for the automatic acquisition of workflow cases from free text.

Figure 2 summarizes the main topics and the commonalities that can be found in the book chapters.

3 Conclusion

This book collects a set of chapters on successful CBR applications. Very interestingly, most of them can be classified as PO-CBR works, testifying the growing interest of the CBR community towards this specific research direction. Indeed, also the non-PO-CBR works have some relations with the topic, specifically as regards case acquisition and case representation.

Some of these contributions propose methodological solutions that, even though applied to a specific domain, could be properly adapted to different applications as well. They demonstrate the capability of CBR to solve or handle issues which would be too difficult to manage with other classical AI methods and techniques.

Other chapters report on frameworks and tools that have now reached a very mature development stage, being of interest not only for academic research, but also for industrial exploitation.

In conclusion, the selected works testify the flexibility of CBR, and its applicability as a decision support methodology in fields (like business process management) where experiential knowledge plays a key role, exceptional situations often take place and have to be properly managed, and past solutions can be fruitfully collected for evaluation and reuse.

References

1. Aamodt, A., Plaza. E.: Case-based reasoning: foundational issues, methodological variations and systems approaches. AI Commun. **7**, 39–59 (1994)
2. Aha, D., Munoz-Avila, H.: Introduction: interactive case-based reasoning. Appl. Intell. **14**, 7–8 (2001)
3. Aha, D., Marling, C., Watson, I.: Case-based reasoning commentaries: introduction. Knowl. Eng. Rev. **20**, 201–202 (2005)

4. Bergmann, R.: Experience Management: Foundations, Development Methodology, and Internet-based Applications. Springer, Berlin (2002)
5. Bichindaritz, I.:Memory Structures and Organization in Case-Based Reasoning. Studies in Computational Intelligence, vol. 73, pp. 175–194. Springer, Berlin (2008)
6. De Mantaras, R.L., McSherry, D., Bridge, D., Leake, D., Smyth, B., Craw, S., Faltings, B., Maher, M.L., Cox, M.T., Forbus, K.: Retrieval, reuse, revision and retention in case-based reasoning. Knowl. Eng. Rev. **20**, 215–240 (2005)
7. Elhadi, M.T.: Using statutes-based IR drive legal CBR. Appl. Artif. Intell. **15**(6), 587–600 (2001)
8. Hullermeier, H.: Case-Based Approximate Reasoning. Springer, Berlin (2007)
9. Jurisica, I., Glasgow, J.: Applications of case-based reasoning in molecular biology. AI Mag. **25**(1), 85–95 (2004)
10. Kolodner, J.L.: Case-based reasoning. Morgan Kaufmann, San Mateo, CA (1993)
11. Leake, D.B., Smyth, B., Wilson, D.C., Yang, Q. (eds): Special issue on maintaining case based reasoning systems. Comput. Intell. **17**(2), 193–398 (2001)
12. Minor, M., Montani, S., Recio Garcia, J. (eds): Special issue on process-oriented case-based reasoning. Inf. Sys. (2013, in press)
13. Montani, S.: How to use contextual knowledge in medical case-based reasoning systems: a survey on very recent trends. Artif. Intell. Med. **51**, 125–131 (2011)
14. Patterson, D., Rooney, N., Galushka, M.: A regression based adaptation strategy for case-based reasoning. In: Dechter, R., Kearns, M., Sutton, R. (eds.) Eighteenth National Conference on Artificial intelligence Edmonton, Alberta, Canada, pp. 87–92. American Association for Artificial Intelligence, Menlo Park, CA (2002)
15. Surma, J., Vanhoof, K.: Integration rules and cases for the classification task. In: Veloso, M., Aamodt, A. (eds) Proceedings of the 1st International Conference on Case-Based Reasoning, Sesimbra, Portugal. Lecture Notes in Computer Science, vol. 1010, pp. 325–334. Springer, Berlin (1995)
16. Watson, I.: Applying Case-Based Reasoning: Techniques for Enterprise Systems. Morgan Kaufmann, San Mateo (1997)
17. Wilson, D.R., Martinez, T.R.: Improved heterogeneous distance functions. J. Artif. Intell. Res. **6**, 1–34 (1997)

Chapter 2
A Study of Two-Phase Retrieval for Process-Oriented Case-Based Reasoning

Joseph Kendall-Morwick and David Leake

Abstract Process-Oriented Case-Based Reasoning (PO-CBR) systems often use structured cases, which in turn require effective structure-based retrieval methods, especially when dealing with large processes and/or large case bases. Good retrieval performance can be facilitated by two-phased retrieval methods which first winnow candidate cases with a comparatively inexpensive retrieval phase, and then apply a more expensive strategy to rank the selected cases. Examples of such processes have been shown to provide good retrieval results in limited retrieval time. However, studies of such methods have focused primarily on overall performance, rather than on how the individual contributions of each phase interact to affect overall performance. This misses an opportunity to tune the component algorithms in light of the system task and case base characteristics, for specific task needs. This chapter examines two-phased retrieval as a means of addressing the complexity in many PO-CBR domains, and specifically examines the performance of each phase of two-phased retrieval individually, demonstrating characteristics of the phases' interaction and providing general lessons for how to design and deploy two-phased retrieval systems.

1 Introduction

Process-oriented tasks, such as business process management, workflow generation, and planning, require generating complex structured solutions. The generation of such solutions may be difficult due to imperfect domain knowledge and computational complexity. Case-based reasoning (CBR) (e.g., [1–3]) is appealing for

J. Kendall-Morwick (✉)
Computer Science Department, DePauw University, Greencastle, IN 46135, USA
e-mail: josephkendallmorwick@depauw.edu

D. Leake
School of Informatics and Computing, Indiana University, Bloomington, IN 47408, USA
e-mail: leake@cs.indiana.edu

S. Montani and L. C. Jain (eds.), *Successful Case-based Reasoning Applications-2*,
Studies in Computational Intelligence 494, DOI: 10.1007/978-3-642-38736-4_2,
© Springer-Verlag Berlin Heidelberg 2014

addressing both problems. Because CBR generates solutions by revising the lessons of relevant prior solutions, it can increase efficiency by providing a head-start to problem-solving. Because its reasoning process has a strong analogical component, it enables solution generation to benefit from useful characteristics of retrieved solutions even if their relevance is not explicitly encoded. Consequently, process-oriented case-based reasoning (PO-CBR) has attracted considerable interest in the CBR community (e.g., [4]), for a wide range of tasks such as assisting in the development of new processes, supporting the enactment of existing processes, or the adaptation of processes during their execution.

A commonality across many PO-CBR tasks is the importance of representing structured information. Such information is generally represented in the form of labeled or attributed graphs, on which graph matching can play an important role in similarity judgments. However, graph matching algorithms have high computational cost, and their performance is constrained by theoretical limits: aspects of the graph matching problem are NP-complete. As a result, heuristic methods may be needed to speed up retrieval, even for small case bases, and especially for interactive applications requiring rapid response time. The need for interactive assistance places a special premium on retrieval efficiency, which may sometimes require balancing efficiency concerns against the quality of case selection when designing retrieval algorithms.

Greedy algorithms are often applied to structured case retrieval, but depending on the complexity of the data, greedy methods may not be optimal choices. Beyond the issue of how to compare a query case to a case in the case base, it may be necessary to focus the choice of which cases should be compared: An exhaustive case-by-case comparison wastes time comparing irrelevant cases to the query, but too strict restrictions on cases to compare could result in overlooking important cases.

This chapter provides a perspective on the retrieval of structured cases, illustrated with examples from the retrieval methods of Phala, a workflow authorship assistant. It focuses especially on two-phased techniques for retrieval, as used by Phala and also explored by others within the sub-field. Tools for supporting two-phase retrieval, developed for Phala, are implemented in a freely available system for structured case retrieval[1] which was used as the basis for the experiments in this chapter.

In particular, the chapter examines the design questions underlying development of two-phased retrieval systems and their ramifications on the performance of the two individual retrieval phases. The overall performance of two-phased retrieval systems has been studied in many task contexts (e.g., [5]). However, little study has been devoted to a key factor in how to design such systems: the individual contributions of the two phases. Analyzing such factors is important because it affects the characteristics to be sought in the design of each phase. This chapter presents experimental studies designed to illuminate the roles of each phase.

Results of the experiments suggest the value of tailoring overall parameter choices for two-phased retrieval choices to case base characteristics: in addition to any domain-specific choices for retrieval processes within each phase, the two-phased

---

[1] The Structure Access Interface (SAI): http://www.cs.indiana.edu/~jmorwick/SAI

retrieval strategy must itself be evaluated and the selectivity of each phase tuned. Our evaluation provides insight into what the important parameters and performance metrics for two-phased retrieval are, how changes in these parameters can affect performance, and how properties of the case base can complicate these relationships.

The chapter begins with background on PO-CBR and case retrieval in CBR. The following section expands by exploring representation and retrieval of structured cases. Section 4 goes into greater detail on two-phased retrieval, discussing various indexing strategies for two-phased retrieval and how to tune a two-phased retrieval configuration for an optimal combination of response time and accuracy, supported by an empirical study of a simple indexing strategy. The evaluation is performed using the Structure Access Interface system (SAI), a retrieval system developed for PO-CBR [6]. The chapter closes with observations and opportunities for future work.

2 Background

2.1 Retrieval in Case-Based Reasoning

The case-based reasoning process is often framed in terms of a cyclical process with four tasks which guide the access and application of stored cases, and learning through case storage: retrieve, reuse, revise, and retain [1]. Many retrieval approaches have been explored (see [3] for an overview). Often retrieval algorithms work in concert with cleverly designed domain-specific indexing schemes, to enable efficient selection of cases expected to be useful (e.g., [7, 8]); sometimes indices are designed to summarize structural characteristics, to enable retrieval to reflect structure without requiring full structure matching (e.g., [9]). However, domain-independent retrieval approaches cannot rely on such indexing obviating the need for structure matching. The work in this chapter focuses primarily on enabling rapid structural matching, which in turn can leverage semantic information when available.

2.2 Process-Oriented Case-Based Reasoning

PO-CBR is concerned with the application of case-based reasoning to process-oriented domains and with the resultant issues and opportunities [10]. PO-CBR research touches on many different domains, for example, business processes, e-Science, cooking, software development, health care, and game development, and PO-CBR systems use cases developed from data for varying types of processes such as workflows, plans, software models, and state machines. PO-CBR also studies the application of CBR to artifacts from processes which have executed, as for traces [11], logs, and provenance [12]. The study of workflows within PO-CBR has mainly fallen into two domains, business processes and e-Science.

Business Processes Business Process Management (BPM) concerns the management of business processes, the steps and procedures taken by an organization as whole to conduct some aspect of its business [13]. These steps are often modeled by workflows, which identify the order and conditions under which each step is taken, and what data or materials, potentially produced in a prior step, is necessary for each step.

BPM has been studied for several decades, and work to support BPM, particularly from CBR researchers, has gained traction in the last decade. Recently, Montani and Leonardi applied PO-CBR to cases of medical processes for the management of strokes, to determine their medical correctness [14]. Minor et al. have developed methods to support digital design by with a case-based means of adapting workflows [15]. Kapetanakis et al. have developed a system, CBR-WIMS, which provides a generic, case-based means of intelligent monitoring of business processes [16].

e-Science Information technology is playing an increasingly critical role in scientific experimentation, particularly in *e-Science*. e-Science consists of *in silico* experimentation (experiments that are run completely through execution of computer programs) as well as computational processes intended for data analysis and knowledge discovery [17]. Workflow technology is often employed in e-Science to manage these processes. Workflows developed for these purposes, called *Scientific Workflows*, consist of references to services (remote web-services, scripts, and local programs) and instructions for controlling the order of their execution and the flow of data between them. These services enact simulations, analysis, or transformations of data.

Some scientific workflows are small, but many contain a hundred steps or more, making them challenging for humans to compose. Also, the number of workflows that a scientist runs may be large. For example, ensemble simulation workflows run hundreds of highly similar workflows, differing slightly in structure or in parameters, to perform a parameter sweep study. In addition, the amount of data produced and consumed by services on the grid can be extremely large, and processing times long, making it important to generate the right workflow on the first attempt to avoid wasting computational resources. This provides the practical motivation for our work to apply PO-CBR to support scientific workflow generation.

The Phala Workflow Authorship Assistant Workflow authors are faced with many choices throughout the workflow generation process. Consequently, a number of efforts have made to assist scientists in workflow authorship, including workflow management systems such as Taverna [18], Kepler [19], Trident [20], and XBaya [21]. Additionally, resources such as myExperiment [22] and Biocatalogue [23] assist workflow authors in searching for past workflows or relevant services to include in their workflows.

Phala is a PO-CBR system that has been developed to assist authors of scientific workflows in the creative process of developing new workflows [24]. Like myExperiment, Phala relies on past works to provide assistance to workflow authors, however Phala differs in that users of the system are not burdened with locating relevant works and extracting the relevant information from those works; instead these are integrated into the Phala system, as part of its realization of the CBR cycle. Phala has been shown to produce on-point recommendations in a leave-one-out evaluation over a set of workflows from myExperiment.

Retrieval in Phala Phala is an interactive assistant, offering recommendations for edits to a partially completed workflow to the workflow author. Workflow authors use their discretion to determine which recommendations should be incorporated into the workflow.

In order to be an effective assistant, Phala must be able to react to user queries in seconds. However, rapid retrieval is made more difficult because Phala uses structured cases to represent dataflow through past execution traces of workflows (provenance)—which could slow retrieval to an unacceptable level for even a moderately sized case-base. The problem is further aggravated because Phala could potentially work with very large case-bases.

3 Structured Cases in CBR

Structured cases include information that is not global within the scope of the case; rather, it depends on relationships between individual components of the case. Graphs are a very common and powerful formalization frequently used to provide structure within data. Particularly useful are labeled graphs, which combine both data and structure. In labeled graphs, nodes are labeled with the information constrained to a particular component (an object or a concept). This information may be more complex than a simple classification. For example, ShapeCBR is a system that assists in the design of metal castings [25]. It represents casting components as values of which there are eight possible classifications. However, other details exist for individual instances of these components, such as scale. Lately, the cooking domain has been popular in CBR research, with recipes as cases [26]. In these cases, nodes may represent cooking tasks such as baking, which may be further parameterized by details such as temperature and duration.

Edges represent relationships between these entities. If only one type of relationship is modeled, edge labels are not necessary. However, this is often not the case. For instance, an unbounded number of relationship types must be modeled in the Concept Map CBR domain [27]. Likewise, in ShapeCBR represent physical linkage between casting components, and are not all equivalent; many components do not have an axis of symmetry between each joining location. Furthermore, different kinds of coupling objects are used to link larger components. This information must also be captured by the edge's labels.

Structured cases also have global attributes, similar to traditional feature vectors. In this sense they are a generalization of feature vectors, but more importantly, the global attributes provide a means to store unstructured data. For instance, in workflow domains, a number of global features exist, including semantic tags, contextual features such as author data, and data describing the structured components, such as the workflow language used.

3.1 Mapping and Comparing Structured Cases

Similarity metrics for structured cases can involve edit distance: the more edits required to transform one case to a state in which it is isomorphic to the other, the greater the difference between the two cases (e.g. [15]). Another, similar approach is to find the maximal common subgraph, which involves finding a correspondence between the nodes in each case and examining the number of resulting corresponding edges (e.g. [6]).

In either approach, determining the compatibility of nodes or edges to form a mapping between components of each case involves determining the compatibility of the features stored in the components. This can be a purely syntactic comparison, or can involve semantic constraints on the similarity of these features, such as those explored by Bergmann et al. [28, 29].

A key issue for retrieval and similarity assessment of structured data is that graph processing algorithms can be computationally expensive (for instance, subgraph isomorphism is NP-Complete [30]). A general retrieval tool must address the inherent complexity in these domains in order to handle similarity queries. Numerous methods have been developed to address these computational constraints, such as greedy algorithms, anytime algorithms, and heuristic search. For instance, Bergmann and Gil have adopted a knowledge intensive A* approach to mapping structured cases to queries in a workflow processing domain [28]. Reichherzer and Leake have used topological features of concept maps to weight the significance of local features [9]. Minor et al. have addressed the complexity in the workflow domain by converting graph structures to flat strings for comparison [31].

4 Two-Phased Case Retrieval

Achieving high retrieval performance requires both speeding up the case comparisons which are done and avoiding costly comparisons wherever possible. This approach has roots in cognitive science [32], as described below, and the idea of retrieving structures in two phases was adopted at an early stage by CBR researchers working with structured data [33, 34].

In a two-phased retrieval process, an initial "coarse-grained" and comparatively inexpensive retrieval process is used to winnow the cases to be considered. A more expensive "fine-grained" strategy is then applied to the pool of candidate cases selected by the coarse-grained strategy. PO-CBR researchers have noted the potential value of this strategy for scaling up case base sizes [35, 36], as have researchers in graph database systems [37–40]. For example, in one comparison, the Phala CBR system required approximately 15 minutes to apply its similarity metric to each case in a moderately sized case-base containing about 350 workflow cases, but required only a few seconds on the same hardware to perform the same search with two-phased retrieval and appropriate indexing for the first phase.

The use of indexing in two-phased retrieval makes it possible to add and remove cases with relatively little maintenance. Two-phase retrieval is flexible in that it can support multiple similarity measures simultaneously, to be applied as needed. It has some potential drawbacks—for example, the recall of the retrieval process will decrease if the coarse-grained strategy misses relevant cases—but for many tasks has proven useful compared to single-phase approaches.

In the remainder of this chapter, we use the following terminology for two-phased retrieval: *Phase 1* refers to the coarse-grained retrieval process aimed at efficiently selecting a subset of cases for more complete consideration, and *Phase 2* to the more computationally complex fine-grained ranking process. The maximum number of cases Phase 1 provides to Phase 2 is the Phase 1 Window, shortened to *Window 1*. The number of cases returned from the Phase 2 process for further processing by the CBR system is called the Phase 2 Window, shortened to *Window 2*.

MAC/FAC Gentner and Forbus's MAC/FAC ("Many are called but few are chosen") model, an early two-phased retrieval approach, was inspired by humans' capability to access structured knowledge quickly and soundly [32]. Gentner and Forbus argue that human retrieval speed implies that human retrieval must employ a fast, coarse-grained, massively parallel process which need not fully reflect structural constraints, and that the soundness implies the existence of a more thorough and precise process, which weighs structural constraints. The MAC/FAC algorithm models this behavior by performing retrieval in two sequential phases. The MAC phase involves comparing candidate cases (structures) through flat representations of features of their structural content. The FAC phase chooses a range of the top ranked cases from the MAC phase and re-orders them according to a more precise and computationally complex similarity metric. The smaller set of the top ranked cases from the FAC phase is returned.

The addition of the FAC phase can dramatically speed up retrieval by using a sublinear time lookup (in terms of the size of the case-base) for each case indexed by their features. Only the cases deemed most likely to be similar will then reach the FAC phase and have their structure examined and mapped in a more costly process (potentially requiring computation of the maximum common subgraph, or other hard graph comparison techniques).

Recent Applications of Two-Phased Retrieval Many systems have more recently applied two-phased retrieval approaches. The use of such approaches is a strong trend in the growing research area of graph databases [41]. Many graph database systems use a two-phased approach in which graphs are indexed by structural features used for an initial coarse-grained retrieval phase. During this phase, structural details of the graph are not loaded into memory, but are instead inferred through comparison of their association with flat indices representing structural summaries. Examples of graph database systems employing such a technique are Graphgrep [40], G-Hash [37], Periscope/GQ [38], and gIndex [39].

The SAI system, developed for Phala, joined this collection as a graph database system specifically tailored to the needs of PO-CBR researchers [42]. SAI provides a flexible means of implementing a retrieval algorithm. SAI was developed to assist

PO-CBR researchers and is intended for use in studying various retrieval algorithms of structured cases applied to various domains.

4.1 Indexing Strategies

Indexing is an effective means for efficient retrieval in large case-bases. Indices can be used to search for the most important and relevant cases without requiring a similarity assessment for each case in the case-base. Indexing is related to similarity assessment in that index selection should yield cases likely to be assessed as more similar. The representational power of a system's indices bears directly on how accurately index-based retrieval can estimate the outcome of a full similarity assessment.

Indices should capture important case properties in order to allow for content-based retrieval methods, such as conversational retrieval, retrieval by semantic tag, context-sensitive retrieval, and others. For example, Weber et al. retrieve related workflow instances through an interactive, conversational [43] process with the user [44], answering questions associated with cases in the case-base and retrieving cases related to the user's responses. Allampalli-Nagaraj et al. use semantic metadata tags to improve retrieval of medical imaging cases in their CBR application [45].

Many global case properties are easily attainable "surface features", stored in a literal form within the case. This is in contrast to deep features, which sometimes must be derived from the case data through potentially computationally expensive means [33]. Structural features are readily apparent in the case data, but they may correspond to deep features, because they are chosen from a very large feature-space (potentially constrained by the type of structural indexing used).

The following subsections detail three approaches to forming structural indices, with various degrees of distinguishing power and limits on the size of the index-space. An important consideration for these approaches is that each has its own advantages, and for a particular domain, any one of these methods may be optimal.

Path-1 Indexing SAI currently includes a simple indexing strategy that enumerates all paths of length 1 (a single edge), including select attributes from the edge and connected nodes. An example of this type of index can be seen in Fig. 1.

This indexing option was developed because the single-edge indices hold the advantage of $O(n)$ index generation while maintaining sufficient distinguishing power within the e-Science workflow domain to support Phala's retrieval needs. These short path indices are sufficient for the e-science domain but not for others, as later sections will demonstrate. This agrees with the result from the iGraph experiment, which determined that no single algorithm is best for every dataset in exact supergraph retrieval. This observation was the impetus behind SAI's modular approach, which allows for the integration of custom indexing strategies.

Path-K Indexing Beyond single edge indices such as those used by Path-1, longer paths could be used which increase the distinguishing power of each index. This also increases the size of the space of potential indices, in that graphs will, in the

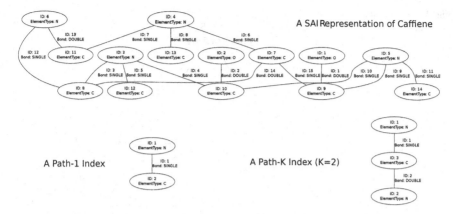

Fig. 1 A SAI representation of caffeine with path indices

worst case, have a $O(n^k)$ number of potential indices. With the growing size of the index-space, it may be advisable to only consider a subset of the space as viable indices. This also complicates the process of determining which viable indices are associated with a query at the time it is presented to the system. A system utilizing this type of indexing strategy is Graphgrep [40]. An example of this type of index can also be seen in Fig. 1.

Arbitrary Substructure Though Path-K indices can increase the distinguishing power of the indices, their topological limitations prevent the indexing strategy from providing optimal results as a phase-1 retrieval strategy for some domains. Yan et al. identify that Path-K indices do not sufficiently represent the structure of a graph in the domain of molecular structures, and present a system for mining common substructures for use in indexing [46]. This is a generalization of the Path-K indexing technique, in which the distinguishing power of indices is further increased, but the number of indices that could be associated with a case is $O(2^n)$ in the worst case scenario. This necessitates selectivity of which potential indices from the larger index-space will be used within a system, requiring a common substructure mining process, and additional complications during query-time to determine which indices are associated with the query. An example of this type of index can be seen in Fig. 2.

4.2 Evaluating Phase 1 Against Phase 2

This section analyzes how each phase affects various performance metrics. The first subsection outlines relevant specifics of the original MAC/FAC approach, the second subsection outlines how this approach can be adapted to address time constraints to accommodate interactive systems, and the third subsection outlines a method for comparing performance of each phase to determine which must be tuned when addressing performance problems.

Fig. 2 An arbitrary substructure index

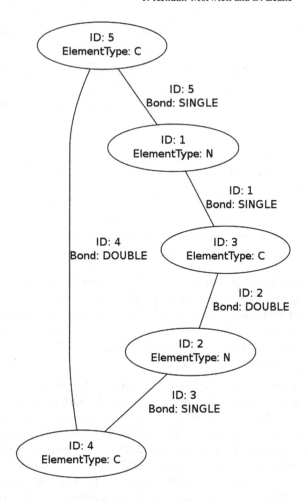

Effects of Limiting Phase 1 Retrieval by Similarity In Phase 1 of the original
MAC/FAC model, the structure in memory sharing the most features with the query
is selected, along with any others matching at least 90 % as many features. In this
sense, the MAC phase works as a filter. Graph databases use a similar approach,
in which features may be mined and used to approximately summarize structural
details. These features are recorded as indices for the cases, enabling Phase 1 to
retrieve a set of potentially relevant cases (cases which share many indices with the
query). For such a Phase 1 process, the number of matches provided for consideration
by Phase 2 depends on two main factors:

- The average size of clusters of cases with high similarity to each other
- The distinguishing power of the features selected for filtering

Applying Phase 2 processing to all cases within 10 % of the best match provides
some assurance as to the quality of the result from Phase 2. For instance, if 99 %

of the time the top-ranked case according to the ranking of Phase 2 has a 90 % or greater match with the query based on Phase 1 features, then Phase 2 will output the top case in the case-base 99 % of the time. Unfortunately, providing this assurance requires that there be no constraint on the number of cases considered by Phase 1, and thus, there is also no lower-bound on the amount of time saved in Phase 2. We can avoid this problem by setting a limit on Window 1.

Using a Fixed Retrieval Window to Bound Retrieval Time If response time is important, as in an interactive system, we can set a bound on retrieval time by limiting the number of cases brought forth from the inexpensive Phase 1 to the significantly more expensive Phase 2. In the MAC/FAC model, cases are judged against each other relative to the degree to which they matched features within the query graph. We note rather than simply using this as a filter, it can be used to generate a ranking. For example, Phase 1 can create a nearest-neighbor ranking of the cases in the case-base according to the similarity of their features to those of the query, with Window 1 set to the maximum number of cases Phase 2 can process within the time allowed for Phase 2. Phase 1 will no longer be returning every case within a given similarity threshold of the query; as a result, the expected response time for a query can be bounded. Window 1 can be adjusted for the desired trade-off between response time and the similarity of the retrieved cases to the query.

Comparing Phase 1 and Phase 2 Rankings for Credit/Blame Assignment Evaluation of two-phased retrieval algorithms has traditionally been performed by examining performance of the system as a whole. The problem with this approach is that, if the desired case is not retrieved, it fails to identify which component of the system is most responsible for failure. Either phase may be responsible: Phase 1 may fail to rank the most relevant cases into the top cases in Window 1, making it impossible for the Phase 2 algorithm to find the most relevant cases. Similarly, Phase 2 may be provided with the most relevant case, but fail to rank it higher than other cases provided from Phase 1. To determine which bears more of the blame, we can compare the rankings from Phase 1 and Phase 2. Assuming that the finer-grained strategy of Phase 2 can be used as a "gold standard," increased difference between the rankings suggests blame for Phase 1. However, comparing similarity rankings, particularly in two-phased retrieval, is less straightforward than might be expected. The following paragraphs examine how such a comparison should be done for each retrieval phase.

Evaluating Phase 1 Retrieval In a study of case ranking for single-phase retrieval, Bogaerts and Leake [47] provide methods for comparing the results of a single-phase retrieval system to an ideal ranking of cases, to evaluate different retrieval algorithms. Adding a second phase to the retrieval process adds complications which may produce unexpected results in isolated cases. However, we have conducted an experiment which suggests that methods which are effective in comparing rankings for single-phased case retrieval systems may also be the most effective for comparing phase 1 and phase 2 rankings. This section first describes the complications which raise questions about assessing rankings for two-phased retrieval, and then sketches our experimental result.

For an example of how the problem of comparing phase 1 and phase 2 rankings can be complicated, suppose that Phase 1 happens to always return the top Window 1 cases of the ideal ranking, but in the opposite order. If Window 1 is large, this ranking scores very low when evaluated for single-phase retrieval. However, the end result of using this ranking for Phase 1 of the two-phased algorithm will be identical to the result of Phase 1 returning all cases in the ideal order: Phase 2 will re-rank the cases, so all that matters is that Phase 2 be provided with the correct set of Window 1 cases.

Another issue, addressed for single-phase systems by Bogaerts and Leake, concerns what they dub "the k-boundary" problem. This problem arises if a set of cases with the same similarity to the query straddle the retrieval window boundary, so that only some of the cases are provided to the next phase. In such situations, obvious methods of assessing retrieval quality may provide counter-intuitive results, making it difficult to determine which ordering is best. Depending on the similarity metrics used, this problem may arise frequently in both phase 1 and phase 2. For instance, we have used a simple matched feature count for phase 1 similarity assessment, which very frequently results in ties. For phase 2, we have used a covered edge count as a similarity metric, which often results in ties for small graphs.

Comparing ordering may also result in other anomalies for two-phased processes. For example, the Phase 1 ranking of the top Window 1 cases in opposite of ideal order will also score lower than some rankings which do not contain all of the top Window 1 cases. In such cases, the lower-scoring ranking will actually provide better performance than the higher scoring ranking! In this sense, traditional rank quality measures are not guaranteed to be a good measure for the effectiveness of Phase 1 ranking in every instance. Figure 3 illustrates such a case when Window 1 is 10 and Window 2 is 5. Ranking 1 will score lower than ranking 2 with a traditional weighted rank measure (our measure rates ranking 1 as 48.4 % similar to the ideal ranking and ranking 2 as 98.0 % similar to the ideal ranking). However, after phase 2 ranking, results from ranking 1 will yield the top 5 cases in order, whereas the results from

Fig. 3 Example rankings

Ideal	Ranking 1	Ranking 2
1	15	1
2	14	2
3	13	3
4	12	4
5	11	6
6	5	7
7	4	8
8	3	9
9	2	10
10	1	11
11	10	5
12	9	12
13	8	13
14	7	14
15	6	15

ranking 2 will miss case 5 and yield case 6 instead. Ranking 1 will yield better results than ranking 2, meaning that the traditional rank measure did not properly identify which ranking was superior.

However, such anomalies are generally unlikely to arise. In the prior example, sufficiently increasing or decreasing Window 1 or Window 2 eliminates the anomaly. In general, if the ranking generated by Phase 1 is a good approximation of the ranking generated by Phase 2, the combined algorithm will be more likely to locate the most similar cases.

To test whether considering rank order when comparing Phase 1 and Phase 2 rankings provides better information about the culpability of Phase 1 versus Phase 2 when the system fails to retrieve desired cases, we conducted experiments with a two-phased retrieval system in which the Phase 1 ranker was identical to the Phase 2 ranker, except for a predetermined random error introduced in Phase 1. We created 2 Phase 1 rankers, each with differing amounts of error. We used two rank measures to compare the rankings generated by each of these Phase 1 rankers: one measure which considered the ordering as part of the quality measure, and one which only considered the presence of the desired cases in the Window 2 set of cases (a recall measure). We ran 1,000 tests on a synthetic dataset, with Window 1 sizes of 20 and 25 and Window 2 sizes 5 and 10, with error differences from 0.01 to 0.30 and base error of 0.1. We then computed rank quality for both Phase 1 rankers with both measures, iteratively, until enough judgments were made to indicate which ranker had a higher error to a statistically significant degree. In all trials, the measure which considered order detected a statistically significant difference between the performance of the two phase 1 rankers faster than the recall measure, and only required an average of 5.5 % of the tests recall required to find a statistically significant difference with a Z test using 99 % confidence intervals. Thus considering the order in which phase 1 ranks cases provided better information on the source of retrieval problems.

4.3 Tuning Two-Phased Retrieval

In this section, we outline a process by which a system designer can answer basic questions about how to implement and deploy a two-phased retrieval subsystem. Specifically, we will examine the trade-offs involved in choosing Window 1 and Window 2 settings. We examined this question by using a simple two-phased retrieval algorithm within the SAI framework and applying it to two different datasets. The test retrieval algorithm indexes cases using the Path-1 indexing strategy which ships with SAI. The structural comparison component is a modified greedy algorithm.

Assessing the Dataset The first step in our process is to assess relevant properties of the case-base itself for prevalence of distinguishing features. Our first test dataset consists of scientific workflows downloaded from the myExperiment website [22]. We have used this dataset to evaluate the Phala CBR system for assisting authorship of scientific workflows [48]. In myExperiment the number of unique features is

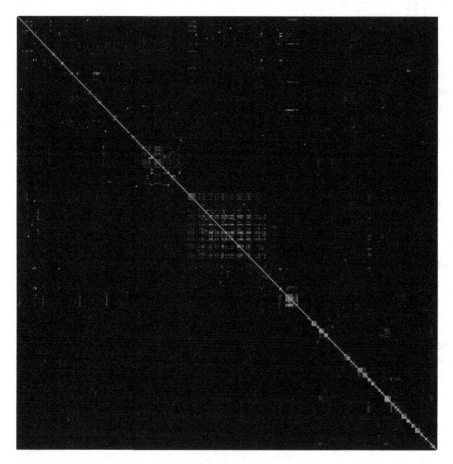

Fig. 4 Pairwise similarity of the myExperiment workflow dataset

relatively large in comparison to the number of nodes within cases in the case-base (specifically, there are 0.65 unique features per node). This results in high performance for our simple retrieval strategy, since the uniqueness of features leads to a larger index space and more distinguishing indices (for 341 graphs, there are 1627 unique indices). Figure 4 is a heat-map illustrating the similarity between each pair of cases from the myExperiment dataset. Note that most of this map is black, reflecting lack of similarity between most case pairs.

Determining Performance Needs The next step is to determine what is needed or desired in terms of performance. Performance of the system can be measured in many ways other than the problem-solving accuracy of the CBR system itself. Response time is important in interactive systems and constraints on the hardware used (e.g., memory limitations) are always relevant. To examine the impact the choice of indexing strategy has for a variety of window settings, and also to examine the

Table 1 Resources used per window 1 size

myExperiment			PubChem		
Window 1	Memory	Time	Window 1	Memory	Time
5	21573	1095	5	20142	5997
8	31768	1688	8	31146	10316
10	37581	1965	10	38811	13296
15	51053	2660	15	54403	19232
20	62741	3130	20	68887	25295
25	71527	3438	25	83618	30688
30	79996	3771	30	96774	35105

effectiveness of our indexing strategy, we generated rankings for both phases with the myExperiment dataset.

The results of this experiment are listed in the left-hand portions of Tables 1 and 2 (and graphed in Fig. 5). Table 1 lists the response time of the entire retrieval process and the memory consumed during retrieval. Response time is indicated in milliseconds and memory is indicated by an implementation-independent unit. Actual memory consumption is a function of this variable which also includes a constant factor for the size of the specific representation of structural elements in a specific implementation. Table 2 indicates how similar the ranking for the top Window 2 cases produced by Phase 1 is to the ranking produced by Phase 2, using a rank measure as described in the previous section.

Table 1 shows that, as Window 1 size increases, memory used and retrieval time also increase. This is expected, but what is more interesting is relating this data to the data in Table 2, which in turn shows that rank quality improves as either Window 2 decreases or Window 1 size increases. Specifically, we note the conflict between these qualities and the trade-off that results. This illustrates that values of Window 1 and Window 2 must be chosen carefully to achieve the optimal result for the system's specific goals.

Table 2 Average rank similarity

myExperiment				PubChem			
Window 1 / Window 2	3	5	10	Window 1 / Window 2	3	5	10
5	88.96	84.23		5	58.20	57.50	
8	92.27	89.01		8	59.83	59.62	
10	93.08	89.98	88.61	10	61.52	61.49	59.91
15	94.34	91.90	90.68	15	65.40	65.70	64.08
20	94.67	92.36	91.21	20	69.86	70.26	68.60
25	94.68	92.46	91.33	25	73.40	73.87	72.17
30	94.97	92.93	91.62	30	77.01	77.49	75.72

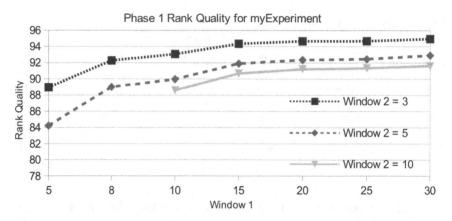

Fig. 5 Phase 1 rank quality with myExperiment dataset

Fig. 6 Pairwise similarity of the PubChem molecular structure dataset

Examining the Impact of the Dataset We used a second dataset to examine how the properties of a dataset bear on decisions for implementing a retrieval system. The second dataset was 100 cases from the PubChem library of chemical structures [49]. Feature variety was much smaller, resulting in 0.0057 features per node and only 8 unique indices. Similarity between cases is much higher, as indicated by the uniformity of bright spots in Fig. 6, a heat-map indicating the similarity between each pair of cases in the PubChem dataset.

Tables 1 and 2 show poor Phase 2 performance compared to that for the first data set, which we ascribe to the small number of indices produced by Path-1. The tables also show the same trends noted for the myExperiment dataset.

5 Conclusions

This chapter presents a broad perspective on retrieval in PO-CBR, including a justification of the need to address complexity issues inherent in retrieval in order to scale up to larger case-base sizes, as well as concrete motivations from our research on case-based methods for supporting generation of scientific workflows. In particular, we have focused on two-phased retrieval to meet these needs.

Through our examination of two-phased retrieval, we have uncovered various factors affecting a trade-off between retrieval accuracy, in terms of the similarity of the cases retrieved, and system responsiveness, in terms of the time taken to respond to a user's query. We have directly examined two main categories of these factors, selection of an indexing strategy, and selection of a window size.

5.1 Selecting an Indexing Strategy

Our experiments illustrate that evaluation of each phase in a two-phased retrieval process may uncover which of the two components is directly contributing to a lack of system performance. In particular, the choice of phase-1 strategy may lead to varied results, depending on the domain. Given that Path-1 offers benefits by eliminating the need to limit the index space by mining frequent structures and also limits the complexity of determining which indices are associated with a query, it appears as a better choice for the myExperiment dataset, considering the rank quality is fairly high. However, this does not carry over to the PubChem dataset. In this case, rank quality is low enough to justify the use of a more complex indexing strategy.

5.2 *Choosing Window Sizes*

Our experiments show that choice of the Window 1 setting can have a dramatic impact
on several competing performance benchmarks, so consideration of this parameter by
a system designer is important to achieve the optimal balance between performance
metrics. An analysis of the selection of the window size with a test dataset, such
as was performed in this chapter, may be useful for determining optimal settings
for a domain, thus liberating the user from the task of fine-tuning this parameter
individually.

6 Future Work

We have outlined a set of considerations for implementing two-phased retrieval,
taking its trade-offs into account and including key questions to examine. We believe
that there are other important questions to ask during this process, and we intend to
elaborate these in future work.

Beyond the feature to node ratio, there are many properties of case-bases which
we expect to correlate with high or low performance for different indexing strategies
(such as the size of the case base and its rate of growth, the average structural similarity
of the cases, and others). We intend to study these characteristics by expanding the
number and type of datasets we use in our experiments.

We also seek to study how these case-base properties predict performance of various indexing strategies. Specifically, we aim to implement the complete Graphgrep
indexing strategy in order to study the effect of index size on the performance metrics for retrieval. We believe examination of choices affecting the size of indices,
the range of index values, or the generality of indices should and will comprise an
additional step in our approach to analyzing and implementing two-phased retrieval
systems.

We seek to expand the performance metrics we consider in this process to include
the time taken to store or index cases and the amount of long-term storage used, as
well as to identify any other key questions not considered within this chapter in our
future work.

We note that two-phased retrieval is not the only alternative for speeding up
structured case retrieval; for example, other viable techniques include clustering
[16], cover trees [50], and fish and shrink [51]. A more thorough comparative analysis
of each of these techniques, which compares the drawbacks and benefits of using
each technique and supports these distinctions with empirical evidence, is warranted.
These techniques should be examined along side two-phased retrieval in an empirical
study to determine strict criteria for judging when one technique is warranted over
another. Additionally, as noted by Bergmann et al. [29], viable alternatives to indexing
exist for performing phase 1 ranking within a two-phased retrieval approach, such as
using surface features or other easily derived features, or techniques such as retrieval

nets [52]. Similarly, a look at such alternatives would be an important aspect of an empirical study.

Acknowledgments This material is based upon work supported by the National Science Foundation under Grant No. OCI-0721674 and by a grant from the Data to Insight Center of Indiana University. Portions of this chapter are adapted from a workshop paper [42] and a dissertation [24].

References

1. Aamodt, A., Plaza, E.: Case-based reasoning: foundational issues, methodological variations, and system approaches. AI Commun. **7**(1), 39–52. http://www.iiia.csic.es/People/enric/AICom.pdf (1994)
2. Leake, D.: CBR in context: the present and future. In: Leake, D. (ed.) Case-Based Reasoning. Experiences, Lessons and Future Directions, pp. 3–30. AAAI Press, Menlo Park (1996)
3. Mantaras, R., McSherry, D., Bridge, D., Leake, D., Smyth, B., Craw, S., Faltings, B., Maher, M., Cox, M., Forbus, K., Keane, M., Aamodt, A., Watson, I.: Retrieval, reuse, revision, and retention in CBR. Knowl. Eng. Rev. **20**(3), 255–260 (2005)
4. Minor, M., Montani, S.: Proceedings of the ICCBR-2012 workshop on provenance-aware case-based reasoning (2012)
5. Han, W.S., Pham, M.D., Lee, J., Kasperovics, R., Yu, J.X.: igraph in action: performance analysis of disk-based graph indexing techniques. In: Proceedings of the International Conference on Management of Data, SIGMOD '11, pp.1241–1242. ACM, New York (2011)
6. Kendall-Morwick, J., Leake, D.: Facilitating representation and retrieval of structured cases: principles and toolkit. Information Systems (2012, in press)
7. Leake, D.: An indexing vocabulary for case-based explanation. In: Proceedings of the 9th National Conference on Artificial Intelligence, pp. 10–15. AAAI Press, Menlo Park (1991)
8. Schank, R., Osgood, R., Brand, M., Burke, R., Domeshek, E., Edelson, D., Ferguson, W., Freed, M., Jona, M., Krulwich, B., Ohmayo, E., Pryor, L.: A content theory of memory indexing. Technical report 1, Institute for the Learning Sciences, Northwestern University (1990)
9. Reichherzer, T., Leake, D.: Towards automatic support for augmenting concept maps with documents. In: Proceedings of 2nd International Conference on Concept Mapping (2006)
10. Minor, M., Montani, S.: Preface. In: Proceedings of the ICCBR-12 Workshop on Process-Oriented Case-Based Reasoning (2012)
11. Floyd, M., Fuchs, B., Gonzlez-Calero, P., Leake, D., Ontañón, S., Plaza, E., Rubin, J.: Preface. In: Proceedings of the ICCBR-12 Workshop on TRUE: Traces for Reusing Users' Experiences—Cases, Episodes, and Stories (2012)
12. Leake, D., Roth-Berghofer, T., Smyth, B., Kendall-Morwick, J. (eds.): Proceedings of the ICCBR-2010 workshop on Provenance-Aware Case-Based Reasoning (2010)
13. Ko, R.K.L.: A computer scientist's introductory guide to business process management (bpm). Crossroads **15**(4), 4:11–4:18 (2009)
14. Montani, S., Leonardi, G.: Retrieval and clustering for business process monitoring: results and improvements. In: Díaz-Agudo, B., Watson, I. (eds.) ICCBR. Lecture Notes in Computer Science, vol 7466, pp. 269–283. Springer, New York (2012)
15. Minor, M., Tartakovski, A., Bergmann, R.: Representation and structure-based similarity assessment for agile workflows. In: Proceedings of the 7th International Conference on Case-Based Reasoning: Case-Based Reasoning Research and Development, ICCBR '07, pp. 224–238. Springer, Berlin (2007)
16. Kapetanakis, S., Petridis, M., Ma, J., Knight, B., Bacon, L.: Enhancing similarity measures and context provision for the intelligent monitoring of business processes in cbr-wims. In: Proceedings of the ICCBR-11 Workshop on Process-Oriented Case-Based Reasoning (2011)

17. Oinn, T., Greenwood, M., Addis, M., Alpdemir, M.N., Ferris, J., Glover, K., Goble, C., Goderis, A., Hull, D., Marvin, D., Li, P., Lord, P., Pocock, M.R., Senger, M., Stevens, R., Wipat, A., Wroe, C.: Taverna: lessons in creating a workflow environment for the life sciences: research articles. concurr. Comput. **18**(10), 1067–1100 (2006)
18. Oinn, T., Addis, M., Ferris, J., Marvin, D., Senger, M., Greenwood, M., Carver, T., Glover, K., Pocock, M.R., Wipat, A., Li, P.: Taverna: a tool for the composition and enactment of bioinformatics workflows. Bioinformatics **20**(17), 3045–3054 (2004)
19. Altintas, I., Berkley, C., Jaeger, E., Jones, M., Ludascher, B., Mock, S.: Kepler: An extensible system for design and execution of scientific workflows. In: Proceedings of the 16th International Conference on Scientific and Statistical Database Management, IEEE Computer Society, Washington (2004)
20. Barga, R., Jackson, J., Araujo, N., Guo, D., Gautam, N., Simmhan, Y.: The trident scientific workflow workbench. In: Proceedings of the 4th IEEE International Conference on eScience, pp. 317–318. IEEE Computer Society, Washington (2008)
21. Shirasuna, S.: A Dynamic Scientific Workflow System for the Web Services Architecture. PhD thesis, Indiana University (2007)
22. Goble, C.A., De Roure, D.C.: Myexperiment: social networking for workflow-using e-scientists. In: Proceedings of the 2nd Workshop on Workflows in Support of Large-Scale Science, WORKS '07, pp. 1–2. ACM, New York (2007)
23. Bhagat, J., Tanoh, F., Nzuobontane, E., Laurent, T., Orlowski, J., Roos, M., Wolstencroft, K., Aleksejevs, S., Stevens, R., Pettifer, S., Lopez, R., Goble, C.A.: Biocatalogue: a universal catalogue of web services for the life sciences. Nucleic Acids Res. **38**(Web-Server-Issue), 689–694 (2010)
24. Kendall-Morwick, J.: Leveraging Structured Cases: Reasoning from Provenance Cases to Support Authorship of Workflows. PhD thesis, Indiana University (2012)
25. Mileman, T., Knight, B., Petridis, M., Preddy, K., Mejasson, P.: Maintenance of a case-base for the retrieval of rotationally symmetric shapes for the design of metal castings. In: Proceedings of the 5th European Workshop on Advances in Case-Based Reasoning, EWCBR '00, pp. 418–430. Springer, London (2000)
26. Stahl, A., Minor, M., Traphöner, R.: Preface: computer cooking contest. In: Schaaf, M. (ed.) ECCBR Workshops, pp. 197–198 (2008)
27. Cañas, A.J., Leake, D.B., Maguitman, A.G.: Combining concept mapping with CBR: towards experience-based support for knowledge modeling. In: Proceedings of the 14th International Florida Artificial Intelligence Research Society Conference, pp. 286–290. AAAI Press, Menlo Calif (2001)
28. Bergmann, R., Gil, Y.: Retrieval of semantic workfows with knowledge intensive similarity measures. In: Proceedings of 19th International Conference on Case-Based Reasoning, Springer, Berlin (2011, in Press)
29. Bergmann, R., Minor, M., Islam, M.S., Schumacher, P., Stromer, A.: Scaling similarity-based retrieval of semantic workflows. In: Proceedings of the ICCBR-12 Workshop on Process-Oriented Case-Based Reasoning (2012)
30. Cook, S.A.: The complexity of theorem-proving procedures. In: Proceedings of the 3rd Annual ACM Symposium on Theory of Computing, STOC '71, pp. 151–158. ACM, New York (1971)
31. Minor, M., Bergmann, R., Grg, S., Walter, K.: Adaptation of cooking instructions following the workflow paradigm. In: Marling, C. (ed.) ICCBR 2010 Workshop Proceedings (2010)
32. Gentner, D., Forbus, K.: MAC/FAC: A model of similarity-based retrieval. In: Proceedings of the 13th Annual Conference of the Cognitive Science Society, pp. 504–509. Cognitive Science Society, Chicago (1991)
33. Mntaras, R.L.D., Bridge, D., Mcsherry, D.: Case-based reasoning: an overview. AI Commun. **10**, 21–29 (1997)
34. Börner, K.: Structural similarity as guidance in case-based design. In: Selected Papers from the 1st European Workshop on Topics in Case-Based Reasoning, EWCBR '93, pp. 197–208. Springer, London (1994)

35. Kendall-Morwick, J., Leake, D.: A toolkit for representation and retrieval of structured cases. In: Proceedings of the ICCBR-11 Workshop on Process-Oriented Case-Based Reasoning (2011)
36. Bergmann, R., Gil, Y.: Retrieval of semantic workflows with knowledge intensive similarity measures. In: Proceedings of the 19th International Conference on Case-Based Reasoning Research and Development, ICCBR'11, pp. 17–31. Springer, Berlin (2011)
37. Wang, X., Smalter, A., Huan, J., Lushington, G.H.: G-hash: towards fast kernel-based similarity search in large graph databases. In: Proceedings of the 12th International Conference on Extending Database Technology: Advances in Database Technology, EDBT '09, pp. 472–480. ACM, New York (2009)
38. Tian, Y., Patel, J.M., Nair, V., Martini, S., Kretzler, M.: Periscope/GQ: a graph querying toolkit. Proc. VLDB Endow. 1, 1404–1407 (2008)
39. Yan, X., Yu, P.S., Han, J.: Graph indexing: a frequent structure-based approach. In: Proceedings of the ACM SIGMOD International Conference on Management of Data, SIGMOD '04, pp. 335–346. ACM, New York (2004)
40. Giugno, R., Shasha, D.: Graphgrep: a fast and universal method for querying graphs. In: Proceedings of 16th International Conference on Pattern Recognition, vol. 2, pp. 112–115 (2002)
41. Angles, R., Gutierrez, C.: Survey of graph database models. ACM Comput. Surv. 40, 1:1–1:39 (2008)
42. Kendall-Morwick, J., Leake, D.: On tuning two-phase retrieval for structured cases. In: Proceedings of the ICCBR-12 Workshop on Process-Oriented Case-Based Reasoning (2012)
43. Aha, D., Breslow, L., Munoz-Avila, H.: Conversational case-based reasoning. Appl. Intell. 14, 9–32 (2001)
44. Weber, B., Reichert, M., Wild, W.: Case-base maintenance for ccbr-based process evolution. In: Roth-Berghofer, T., Göker, M.H., Güvenir, H.A. (eds.) ECCBR. Lecture Notes in Computer Science, vol. 4106, pp. 106–120. Springer, Berlin (2006)
45. Allampalli-Nagaraj, G., Bichindaritz, I.: Automatic semantic indexing of medical images using a web ontology language for case-based image retrieval. Eng. Appl. Artif. Intell. 22(1), 18–25 (2009)
46. Yan, X., Yu, P.S., Han, J.: Substructure similarity search in graph databases. In: Proceedings of the ACM SIGMOD International Conference on Management of Data, SIGMOD '05, pp. 766–777. ACM, New York (2005)
47. Bogaerts, S., Leake, D.: Formal and experimental foundations of a new rank quality measure. In: Proceedings of the 9th European conference on Advances in Case-Based Reasoning, ECCBR '08, pp. 74–88. Springer, Berlin (2008)
48. Leake, D., Kendall-Morwick, J.: Towards case-based support for e-science workflow generation by mining provenance. In: Proceedings of the 9th European Conference on Advances in Case-Based Reasoning, ECCBR '08, pp. 269–283. Springer, Berlin (2008)
49. Bolton, E.E., Wang, Y., Thiessen, P.A., Bryant, S.H.: PubChem: integrated platform of small molecules and biological activities. Annu. Rep. Comput. Chem. 4, 217–241 (2008)
50. Beygelzimer, A., Kakade, S., Langford, J.: Cover trees for nearest neighbor. In: Proceedings of the 23rd International Conference on Machine Learning, ICML '06, pp. 97–104. ACM, New York (2006)
51. Schaaf, J.W.: Fish and shrink. a next step towards efficient case retrieval in large-scale case bases. In: Proceedings of the 3rd European Workshop on Advances in Case-Based Reasoning, EWCBR '96, pp. 362–376. Springer, London (1996)
52. Lenz, M., Burkhard, H.D.: Case retrieval nets: basic ideas and extensions. In: Grz, G., Hldobler, S. (eds.) KI-96: Advances in Artificial Intelligence. Lecture Notes in Computer Science, vol. 1137, pp. 227–239. Springer, Berlin (1996)

Chapter 3
Non-exhaustive Trace Retrieval for Managing Stroke Patients

S. Montani and G. Leonardi

Abstract Retrieving and inspecting traces that log medical processes execution can be a significant help in exception management, and is the first step towards a thorough analysis of the service provided by an health care organization. In this work, we report on extensive retrieval experiments, conducted on a database of 2000 real patient traces, collected at different stroke management units in the Lombardia region, Italy. In our approach, retrieval exploits a K-Nearest Neighbor technique, and relies on a distance definition able to explicitly take into account temporal information in traces—since in emergency medicine the role of time is central. Retrieval is also made faster by the application of non-exhaustive search procedures, that are described in the paper.

1 Introduction

Health care organizations are increasingly facing pressure to reduce costs, while, at the same time, improving the quality of care. In order to reach such a goal, expert physicians (and health care administrators) need to optimize the patient care processes they implement, also by adapting them in front of exceptions (i.e. expected or unanticipated changes and problems in the operating environment [14]), and need to continuously evaluate the services the institution provides.

Patient care processes are usually automated and logged by means of the workflow technology, and therefore recorded as **traces** of execution. A trace is the sequence of the actions that were actually performed, often coupled with their starting and

S. Montani (✉)
DISIT, Computer Science Institute, Universita' del Piemonte Orientale,
Viale Michel 11, 15121 Alessandria, Italy
e-mail: stefania.montani@unipmn.it

G. Leonardi
DISIT, Sezione di Informatica, Università del Piemonte Orientale, Alessandria, Italy

S. Montani and L. C. Jain (eds.), *Successful Case-based Reasoning Applications-2*,
Studies in Computational Intelligence 494, DOI: 10.1007/978-3-642-38736-4_3,
© Springer-Verlag Berlin Heidelberg 2014

ending time. In the simplest form, a trace does not log any other feature about the executed actions (e.g. the actor, or the available resources).

Retrieving and inspecting past traces which are similar to a given process instance, being currently applied to a specific patient, can support physicians if the current process execution entails an atypical situation. Indeed, suggestions on how to adjust the default patient care schema in the current case may be obtained by analyzing the most similar retrieved examples of change, recorded as traces that share the starting sequence of actions with the input problem.

Moreover, a detailed inspection of the retrieved traces can be a first step in the complex process of quality of service evaluation, as it can help in highlighting anomalous situations, before further analyses take place. For instance, some traces may be identified as anomalous (i.e. not similar to the majority of the other, more standard ones), and may therefore need to undergo a formal (e.g. logic-based) verification of compliance with respect to specific semantic constraints.

In this paper, we describe a *case-based retrieval* tool [1], where cases are represented as traces, specifically developed for the stroke management domain.

A stroke is the rapidly developing loss of brain function(s) due to disturbance in the blood supply to the brain. It can cause permanent neurological damage, complications, and death. It is the leading cause of adult disability in the United States and Europe, the number two cause of death worldwide, and may soon become the leading one. Given the severity and the social impact of such a pathology, a very high quality of stroke care is obviously mandatory.

Being stroke a medical emergency, the role of time in care provision is central (at least in the initial stages of patient management). Therefore, in our approach retrieval exploits a proper metric, in which temporal information in traces is explicitly dealt with.

Since retrieval time can become computationally expensive when working on very large databases (as highlighted in e.g. [3]), we have also implemented a methodology able to enhance the performance of our tool, avoiding exhaustive search of similar traces. Specifically, we are resorting to a *Pivoting-Based Retrieval* (PBR—see e.g., [27, 31]) technique, which allows one to focus on particularly promising regions of the search space, and to neglect the others.

The paper is structured as follows: in Sect. 2 we describe stroke management traces, and we provide our metric definition. In Sect. 3 we detail the non-exhaustive retrieval procedure. In Sect. 4 we report on experimental results. Section 5 is devoted to related works, and Sect. 6 addresses our concluding remarks.

2 A Metric for Stroke Management Traces

In our application, we deal with traces logging examples of stroke patients care. A **trace** is a sequence of actions, each one stored with its execution starting and ending times. Additional action features (e.g. actors, available resources) are not recorded in the current version of our framework. Therefore, an action is basically just a

Fig. 1 A view of (part of) five stroke management traces. Identical actions are depicted in the same color (e.g. *yellow actions* are Computed Assisted Tomography), but can have a different duration (e.g. Computed Assisted Tomography lasts longer in the first trace). Examples of different qualitative constraints are provided as well. For instance, *pink* and *purple actions* overlap in trace 1, while *pink action* is before *purple action* in trace 2 (see *round shapes*). On the other hand, *red action* is before *orange action* both in trace 2 and in trace 5, but the length of the delay in between them is different in the two cases (see *square shapes*)

symbol (plus the temporal information). The only type of control flow structure we explicitly deal with is *sequence*, since traces record what has already been executed. This means that alternatives are not possible, and iterations are completely unfolded. Partial or complete parallelism between actions may have taken place, and can be derived from action starting and ending times. Indeed, starting and ending times allow to get information about action durations, as well as qualitative (e.g. Allen's *before*, *overlaps*, *equals* etc. [2]) and quantitative temporal constraints (e.g. delay length, overlap length [18]) between pairs of consecutive actions.

Figure 1 provides a view of (part of) five stroke management traces.

Given the types of information we can find on traces (and the key role of time in the stroke domain), we have introduced a metric to compare them, that takes into account atemporal information (i.e. action types), as well as temporal information (i.e. action durations, qualitative and quantitative constraints between pairs of consecutive actions). Indeed, when evaluating the distance between two traces, it is mandatory to penalize the fact that the very same action had different durations, or was delayed in one case (in fact, anomalous action lengths and delays have to be justified, e.g. for legal purposes, and sometimes may even be life threatening for patients).

Operatively, we first take into account **atemporal information**, by calculating a modified edit distance which we have called **Trace Edit Distance**.

As the classical edit distance [19], our metric tests all possible combinations of editing operations that could transform one trace into the other one. Then, it takes the combination associated to the minimal cost. Such a choice corresponds to a specific alignment of the two traces, in which each action in one trace has been matched to an action in the other trace—or to a gap. We will call it the *optimal alignment* henceforth.

Technically, in order to calculate Trace Edit Distance, we consider edit operations (i.e. substitutions, insertions and deletions) on actions, which are treated as symbols. In particular, in our approach, the cost of a *substitution* is not always set to 1, as in the classical edit distance [19]. In fact, we define it as a value $\in [0, 1]$ which depends on what action appears in a trace as a substitution of the corresponding action in the other trace. In particular, we organize actions in a *taxonomy*, on the basis of domain knowledge. The closer two actions are in the taxonomy, the less penalty has to be introduced for substitution ([26]; see also [4, 25, 29]).

In detail, in our work substitution penalty is set to the *Taxonomic Distance* between the two actions [26], i.e. to the normalized number of arcs on the path between the two actions in the taxonomy:

Definition 1 Taxonomic Distance.
Let α and β be two actions in the taxonomy t, and let γ be the closest common ancestor of α and β. The *Taxonomic Distance* $dt(\alpha, \beta)$ between α and β is defined as:

$$dt(\alpha, \beta) = \frac{N_1 + N_2}{N_1 + N_2 + 2 * N_3}$$

where N_1 is the number of arcs in the path from α and γ in t, N_2 is the number of arcs in the path from β and γ, and N_3 is the number of arcs in the path from the taxonomy root and γ.

The Trace Edit Distance $trace_{NGLD}(P, Q)$ is finally calculated as the Normalized Generalized Levenshtein Distance (NGLD) [35] between two traces P and Q (interpreted as two strings of symbols). Formally, we provide the following definitions:

Definition 2 Trace Generalized Levenshtein Distance.
Let P and Q be two traces of actions, and let α and β be two actions. The Trace Generalized Levenshtein Distance $trace_{GLD}(P, Q)$ between P and Q is defined as:

$$trace_{GLD}(P, Q) = min \left\{ \sum_{i=1}^{k} c(e_i) \right\}$$

where $(e_1, \ldots, e_k)$ transforms P into Q, and:

- $c(e_i) = 1$, if e_i is an action insertion or deletion;
- $c(e_i) = dt(\alpha, \beta)$, if e_i is the substitution of α (appearing in P) with β (appearing in Q), with $dt(\alpha, \beta)$ defined as in Definition 1.

As already observed, the minimization of the sum of the editing costs allows to find the optimal alignment between the two traces being compared.

Definition 3 Trace Edit Distance (Trace Normalized Generalized Levenshtein Distance).
Let P and Q be two traces of actions, and let $trace_{GLD}(P, Q)$ be defined as in Definition 2. We define Trace Edit Distance $trace_{NGLD}(P, Q)$ between P and Q as:

$$trace_{NGLD}(P, Q) = \frac{2 * trace_{GLD}(P, Q)}{|P| + |Q| + trace_{GLD}(P, Q)}$$

where $|P|$ and $|Q|$ are the lengths (i.e. the number of actions) of P and Q respectively.

$trace_{NGLD}(P, Q)$ is just an application of NGLD [35] to traces. Interestingly, it has been proved [35] that NGLD is a metric. In particular, unlike other definitions of normalized edit distance (e.g. [22]), it also preserves the triangle inequality.

Like every variant of the classical edit distance described in the literature, $trace_{NGLD}(P, Q)$ can be calculated resorting to a dynamic programming approach, making its complexity tractable [35].

Once the Trace Edit Distance has calculated the optimal alignment, we resort to it to take into account **temporal information**.

In particular, we compare the durations of aligned actions (e.g. the yellow actions in traces 1 and 3, see Fig. 1) by means of a metric we called **Interval Distance**, defined as follows:

Definition 4 Interval Distance.
Let i and j be two intervals of duration dur_i and dur_j respectively, and let $maxdur$ be the duration of the longest interval available in our trace database. The Interval Distance $interval_d(i, j)$ between i and j is defined as:

$$interval_d(i, j) = \frac{|dur_i - dur_j|}{maxdur}$$

Interval Distance is also exploited to compare the durations of intervals in between aligned actions in the traces (see Fig. 1—square shapes), or the durations of two corresponding overlaps. In the case of delay comparisons, $maxdur$ is set to the duration of the longest delay logged in our trace database.

Once Interval Distance has been calculated referring to all the actions in the two traces being compared, the obtained contributions are summed up. Finally, we divide by the length of the longest trace in the database (in terms of number of actions).

Given such a definition, it is straightforward to prove that Interval Distance is a metric.

When comparing two pairs of corresponding actions, it may however happen that they don't share the same qualitative constraint (see Fig. 1—round shapes). In this case, we cannot resort to Interval Distance to compare the inter-action constraints, because they have a different semantic meaning. On the other end, we can quantify the difference between the two qualitative constraints by resorting to the *A-neighbors graph* proposed by Freska [11] (see Fig. 2).

On such a graph, we can define the Neighbors-graph Distance, as follows:

Definition 5 Neighbors-graph Distance.
Let i and j be two Allen's temporal relations [2], and let G be the A-neighbors graph in Fig. 2. The *Neighbors-graph Distance $ngraph_d(i, j)$* between i and j is defined as:

$$ngraph_d(i, j) = \frac{path(i, j, G)}{max_{k,l \in G}\{(path(k, l, G))\}}$$

Fig. 2 The *A-neighbors graph* proposed by Freska [11]

b=before
m=meets
o=overlaps
s=starts
d-during
f-finishes
e=equals
i*=inverse relations

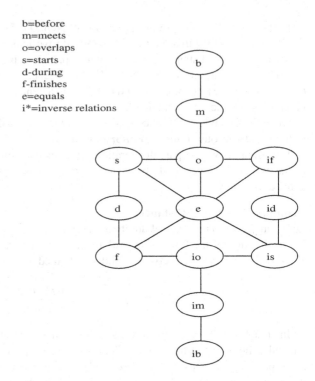

where $path(i, j, G)$ measures the shortest path on G between i and j, and $max_{k,l \in G}\{(path(k, l, G))\}$ normalizes the distance considering the longest path on G.

As above, once Neighbors-graph Distance has been calculated referring to all the actions in the two traces being compared, the obtained contributions are summed up. Finally, we divide by the length of the longest trace in the database (in terms of number of actions).

Given such a definition, it is straightforward to prove that Neighbors-graph Distance is a metric.

Finally, we take the weighted average of Trace Edit Distance, Interval Distance and Neighbors-graph Distance—which is then a metric as well.

3 Non-Exhaustive Retrieval

Our framework exploits the metric defined in Sect. 2 to implement a classical methodology for retrieval, very frequently applied in the first step of Case-Based Reasoning [1] systems, namely **K-Nearest Neighbor (K-NN) retrieval**. K-NN consists in identifying the closest k traces with respect to an input one, according to the distance definition we have introduced, where, of course, the choice of a proper k has to be experimentally set.

As observed in Sect. 2, distance calculation is tractable; nonetheless, it can become computationally expensive when working on very large databases.

Therefore, we have implemented a methodology able to enhance the performance of our tool, avoiding exhaustive search of similar traces. Specifically, we are resorting to **Pivoting-Based Retrieval** (PBR—see e.g., [27, 31]), which allows one to focus on particularly promising regions of the search space, and to neglect the others.

The main idea in PBR consists in:

- computing the distance between a representative case (Pivot) and all the other cases (off-line);
- computing the distance between the Pivot and the input case;
- estimating the distance between the input case and all the remaining cases by using triangle inequality, thus finding a lower and an upper bound for the distance value.

The intervals whose lower bound is higher than the minimum of all the upper bounds can be pruned (see Fig. 3).

Operatively, we initialize $BESTp = \infty$ and $SOL = \{ \}$. We then apply the following iterative procedure:

1. Initialization: $BESTp = \infty$ e $SOL = \{ \}$
2. Choose the Pivot case as the minimum of the midpoints of the intervals; compute the distance between the input case and the Pivot ($DIST$); set $BEST = DIST$;
3. If $BESTp > BEST$ set $SOL = PIVOT$ and $BESTp = BEST$
4. Else if $BESTp = BEST$ set $SOL = \{PIVOT, SOL\}$
5. Prune the intervals whose lower bound is bigger than $BEST$, and remove the Pivot from the set of cases (see Fig. 3)
6. Back to step 2.

We have made tests by defining the Pivot as the mean case, i.e., the one whose average dissimilarity to all the objects in the database is minimal (see Sect. 4). Other choices based on heuristics can be considered as well.

We have also worked at a further approach to non-exhaustive search, in which we first cluster the available traces (resorting to the well-known K-Medoids algorithm [16]), and then select one Pivot for each cluster (i.e., the cluster mean). Specifically, in this case we perform a **two-step non-exhaustive retrieval**, in which:

Fig. 3 Bound pruning in PBR

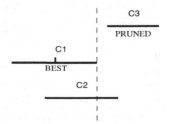

- we identify the cluster the input case should be assigned to;
- we apply the PBR procedure described above to the cluster at hand (taking its mean as the initial Pivot).

We tested both **one-step PBR** (i.e. without clustering), and **two-step PBR** in the stroke domain. Results are reported in the following section.

4 Experimental Results

The quality of retrieval results when adopting our metric was already successfully experimented. Results can be found in [24].

In this paper, we will report on experiments about the non-exhaustive search procedures described in Sect. 3.

Tables 1 and 2 report on our experiments on the use of **one-step** and **two-step PBR** to speed up retrieval time, respectively. We made tests on different case base dimensions (from 250 to 2000 traces; traces were randomly chosen to this end). On every case base, we executed 50 queries. Table 1 compares the average query answering time for retrieving the best 20 cases, without PBR (column 2), and with one-step PBR (column 3). It also provides the average number of pruned cases when resorting to one-step PBR (column 4). Table 2 compares the average query answering time for retrieving the best 20 cases, without PBR (column 2), and with two-step PBR (column 3). It also provides the average number of pruned cases when resorting to two-step PBR (column 4). Finally, column five reports on missed cases, i.e. similar cases that were not retrieved by two-step PBR, because they belonged to clusters excluded in the first step of the method. Experiments were performed on an Intel Core 2 Duo T9400, equipped with 4 Gb of DDR2 ram. Times are in milliseconds.

As it can be observed, retrieval time always improved when resorting to non-exhaustive techniques (see Fig. 4), and up to 40% of the original traces could be pruned in some situations (see Fig. 5). The number of missed cases in two-step PBR was always very small (see Fig. 5), and thus acceptable, according the opinion of the physicians working with us. This consideration is, however, strongly domain dependent (see Sect. 6).

Table 1 Average query answering time (in milliseconds) for retrieving the best 20 cases, without PBR (column 2), and with one-step PBR (column 3) on 5 case bases of growing dimensions

DB dimension	Time no PBR	Time 1-step PBR	Pruned 1-step
250	93.72	70.74	67.28
500	194.86	145.02	148.02
1000	355.22	252.58	322.18
1500	556.18	378.84	433.649
2000	672.6	478.34	479.16

The average number of pruned cases is reported in column 4

Table 2 Average query answering time (in milliseconds) for retrieving the best 20 cases, without PBR (column 2), and with two-step PBR (column 3) on 5 case bases of growing dimensions

DB dimension	Time no PBR	Time 2-step PBR	Pruned 2-step	Missed
250	93.72	56.34	104	4.48
500	194.86	116.94	197	6.9
1000	355.22	216.64	331.9	11.32
1500	556.18	330.9	483.3	14.86
2000	672.6	414.82	499.9	16.74

The average number of pruned cases is reported in column 4, while the number of missed cases is reported in column 5

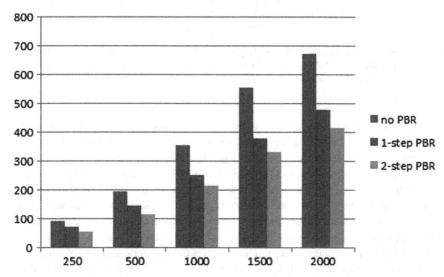

Fig. 4 Comparison between the average query answering time for retrieving the best 20 cases, without PBR, with one-step PBR and with two-step PBR, on 5 case bases of growing dimensions—see Tables 1 and 2 for numerical details

5 Related Work

Since the main methodological contribution of our work consists in the definition of a proper metric, in our comparison with the existing literature we will first focus on this issue, and on the papers providing interesting solutions in relation to it. We will then move to discuss non-exhaustive retrieval, and health care applications.

A number of distance measure definitions for workflows exist. However, these definitions typically require further information in addition to the workflow structure, such as semantic annotations [32], or conversational knowledge [33, 34]. Such approaches are usually context-aware, that is, the contextual information is considered as a part of the similarity assessment of workflows. Unfortunately, any contextual information, as well as conversational knowledge, is not always available, especially

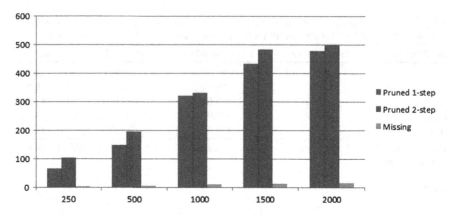

Fig. 5 Number of pruned cases with one-step PBR and with two-step PBR, on 5 case bases of growing dimensions—see Tables 1 and 2 for numerical details. The figure also shows the number of missed cases when applying two-step PBR

when instances of process execution are recorded as traces of actions. Starting from this observation, a rather simple graph edit distance measure [8] has been proposed and adapted for similarity assessment in workflow change reuse [23].

Our approach somehow moves from the same graph edit distance definition. However, with respect to the work in [23], by focusing just on traces of execution we do not need to deal with control flow elements (such as alternatives and iterations). As a matter of fact, traces are always linear, i.e. they just admit the sequence control flow element. From this point of view, our approach is thus simpler than the one in [23].

When focusing on linear traces our approach is more general and flexible. Indeed, we resort to taxonomic knowledge for comparing pairs of actions, so that two different actions do not always have a zero similarity. Moreover, we have introduced a definition which also allows to take into account qualitative and quantitative temporal constraints between actions in process logs. Such a capability is not provided at all in [23].

A treatment of temporal information in trace distance calculation has been proposed in [15]. Somehow similarly to our approach, the metric defined in that work combines a contribution related to action similarity, and a contribution related to delays between actions. As regards the temporal component, in particular, it relies on an interval distance definition which is quite similar to ours. Differently from what we do, however, the distance function in [15] does not exploit action duration, and does not rely on taxonomical information about actions, as we do. Finally, it does not deal with different types of qualitative temporal constraints, since it cannot manage (partially) overlapping actions. We thus believe that our approach is potentially more flexible in practice.

Another contribution [10] addresses the problem of defining a similarity measure able to treat temporal information, and is specifically designed for clinical workflow traces. Interestingly, the authors consider qualitative temporal constraints between matched pairs of actions, resorting to the A-neighbors graph proposed by Freska [11], as we do. However, in [10] the alignment problem is strongly simplified, as they only match actions with the same name. Our approach thus extends their work.

As for the optimization of retrieval, [3] has recently highlighted the opportunity of avoiding exhaustive search in process instance databases, in order to contain computation time. The authors propose an approach to workflow retrieval, where workflow cases are represented as semantically labeled graphs. Some procedures for non exhaustive search, based on the A* algorithm, are provided. However, the approach is not tailored to work on traces, and does not afford the alignment problem, as we do by applying the metric described in Sect. 2.

The work in [17] proposes a two-phase method for process retrieval, which combines an initial and comparatively in-expensive retrieval step, to winnow the cases to be considered, with a more expensive strategy that ranks the remaining cases (as in the well known MAC/FAC system [12]). The paper examines the performance of the two individual phases of two-phase retrieval, demonstrating characteristics of their interaction and providing general lessons for how to design and deploy two-phase retrieval systems. The work in [17] is specifically designed to work on structured cases, however we plan to further analyze it in the future, in order to test it on traces as well.

As a future research direction, we also plan to work on how to apply a retrieval method based on cover trees [5] to our domain. Indeed, such a method can potentially lead to further significant computational improvements.

Interestingly, process management and process analysis issues have gained particular attention in health care applications in the latest years. Just to cite very recent works, in [9] the complexities of health care processes, which are human centric, and multi-disciplinary in nature, are well analyzed. In [30] a system able to support adaptations of running health care process instances is described. A lot of attention is also devoted to conformance checking techniques of clinical guidelines and pathways (see e.g. [7, 13]). The works in [6, 20, 21, 28] deal with process mining in health care. Specifically, [20, 21] apply process mining techniques to the stroke management and to the gynecological oncology domains, respectively. The contributions in [6, 21, 28] also propose pre-processing techniques to improve process mining (which are highly recommended in the medical domain, due to its complexity, as discussed in the already cited [9]).

In summary, the panorama on process management and analysis in health care, the domain to which we are applying our framework, is huge and very active. However, it is worth noting that our work is absolutely general. Indeed, we plan to test it in non-medical domains as well.

6 Concluding Remarks and Future Work

In this paper, we have described a metric for trace retrieval, whose most significant feature is the capability of properly taking into account temporal information in traces. Moreover, a non-exhaustive search strategy has been presented, to make retrieval faster.

We have obtained very encouraging results when applying our technique in the field of stroke management. Indeed, in our experiments retrieval time always improved, especially when resorting to two-step PBR.

In the future, we will also evaluate additional retrieval optimization strategies, in order to contain retrieval time even more. To this end, our research will move along the lines described in [17] and in [5]. We believe that these enhancements really have the potential to make our tool more effective.

In the experiments, we also found that two-step PBR did not lead to a significant loss of similar cases. However, when applying the clustering step, the expert's opinion should always be considered, to carefully evaluate the trade-off between the computational advantages of early pruning, and the risk of loosing close neighbors of the input case, since the situation may be different from domain to domain. Tests in other domains are in fact foreseen in our future work, in order to prove our tool's general applicability.

References

1. Aamodt, A., Plaza, E.: Case-based reasoning: foundational issues, methodological variations and systems approaches. AI Commun. **7**, 39–59 (1994)
2. Allen, J.F.: Towards a general theory of action and time. Artif. Intell. **23**, 123–154 (1984)
3. Bergmann, R., Gil, Y.: Retrieval of semantic workflows with knowledge intensive similarity measures. In: Ram, A., Wiratunga, N. (eds.) Proceedings of International Conference on Case-Based Reasoning (ICCBR), 2011. Lecture Notes in Artificial Intelligence, vol. 6880. Springer, Berlin (2011)
4. Bergmann, R., Stahl, A.: Similarity measures for object-oriented case representations. In: Smyth, B., Cunningham, P. (eds.) Proceedings of European Workshop on Case-Based Reasoning (EWCBR), 1998. Lecture Notes in Artificial Intelligence, vol. 1488. Springer, Berlin (1998)
5. Beygelzimer, A., Kakade, S., Langford, J.: Cover trees for nearest neighbor. In: Proceedings of International Conference on Machine learning, pp. 97–104. ACM, New York (2006)
6. Bose, R.P.J.C., Van der Aalst, W.: Analysis of patient treatment procedures. In: Proceedings of Business Process Management Workshops, vol. 1, pp. 165–166 (2012)
7. Bottrighi, A., Chesani, F., Mello, P., Montali, M., Montani, S., Terenziani, P.: Conformance checking of executed clinical guidelines in presence of basic medical knowledge. In: Proceedings of Business Process Management Workshops, vol. 2, pp. 200–211 (2012)
8. Bunke, H., Messmer, B.T.: Similarity measures for structured representations. In: Proceedings of the European Workshop on Case-Based Reasoning (EWCBR). LNCS, vol. 837, pp. 106–118, Kaiserslautern (1993)
9. Caron, F., Vanthienen, J., De Weerdt, J., Baesens, B.: Advanced care-flow mining analysis. In: Proceedings of Business Process Management Workshops, vol. 1, pp. 167–168 (2012)

10. Combi, C., Gozzi, M., Oliboni, B., Juarez, J.M., Marin, R.: Temporal similarity measures for querying clinical workflows. Artif. Intell. Med. **46**, 37–54 (2009)
11. Freska, C.: Temporal reasoning based on semi-intervals. Artif. Intell. **54**, 199–227 (1992)
12. Gentner, D., Forbus, K.: Mac/fac: a model of similarity-based retrieval. In: Proceedings of Annual Conference of the Cognitive Science Society, pp. 504–509, Cognitive Science Society (1991)
13. Grando, M.A., Van der Aalst, W., Mans, R.S.: Reusing a declarative specification to check the conformance of different cigs. In: Proceedings of Business Process Management Workshops, vol. 2, pp. 188–199 (2012)
14. Heimann, P., Joeris, G., Krapp, C., Westfechtel, B.: Dynamite: dynamic task nets for software process management. In: Proceedings International Conference of Software Engineering, pp. 331–341, Berlin (1996)
15. Kapetanakis, S., Petridis, M., Knight, B., Ma, J., Bacon, L.: A case based reasoning approach for the monitoring of business workflows. In: Bichindaritz, I., Montani, S. (eds.) Proceedings of International Conference on Case Based Reasoning (ICCBR), pp. 390–405. Springer, Berlin (2010)
16. Kaufman, L., Rousseeuw, P.J.: Clustering by means of medoids. In: Dodge, Y. (ed.) Statistical Data Analysis Based on the L1-Norm and Related Methods, pp. 405–416. Elsevier, North-Holland (1987)
17. Kendall-Morwick, J., Leake, D.: On tuning two-phase retrieval for structured cases. In: Lamontagne, L., Recio-García, J.A. (eds.) Proceedings of ICCBR 2012 Workshops, pp. 25–334 (2012)
18. Lanz, A., Weber, B., Reichert, M.: Workflow time patterns for process-aware information systems. In: Proceedings of BMMDS/EMMSAD, pp. 94–107 (2010)
19. Levenshtein, A.: Binary codes capable of correcting deletions, insertions and reversals. Sov. Phys. Dokl. **10**, 707–710 (1966)
20. Mans, R., Schonenberg, H., Leonardi, G., Panzarasa, S., Cavallini, A., Quaglini, S., Van der Aalst, W.: Process mining techniques: an application to stroke care. In: Andersen, S., Klein, G.O., Schulz, S., Aarts, J. (eds.) Proceedings of MIE, Studies in Health Technology and Informatics, vol. 136, pp. 573–578. IOS Press, Amsterdam (2008)
21. Mans, R., Schonenberg, H., Song, M., Van der Aalst, W., Bakker, P.: Application of process mining in healthcare—a case study in a dutch hospital. In: Biomedical Engineering Systems and Technologies, Communications in Computer and Information Science, vol. 25, pp. 425–438. Springer, Berlin (2009)
22. Marzal, A., Vidal, E.: Computation of normalized edit distance and applications. IEEE Trans. Pattern Anal. Mach. Intell. **15**, 926–932 (1993)
23. Minor, M., Tartakovski, A., Schmalen, D., Bergmann, R.: Agile workflow technology and case-based change reuse for long-term processes. Int. J. Intell. Inf. Technol. **4**(1), 80–98 (2008)
24. Montani, S., Leonardi, G.: Retrieval and clustering for business process monitoring: results and improvements. In: Diaz-Agudo, B., Watson, I. (eds.) Proceedings of International Conference on Case-Based Reasoning (ICCBR), 2012. Lecture Notes in Artificial Intelligence, vol. 7466, pp. 269–283. Springer, Berlin (2012)
25. Page, R., Holmes, M.: Molecular Evolution: A Phylogenetic Approach. Wiley, Chichester (1998)
26. Palmer, M., Wu, Z.: Verb semantics for English-Chinese translation. Mach. Transl. **10**, 59–92 (1995)
27. Portinale, L., Torasso, P., Magro, D.: Selecting most adaptable diagnostic solutions through pivoting-based retrieval. In: Leake, D., Plaza, E. (eds.) Proceedings of 2nd International Conference on Case-Based Reasoning. Lecture Notes in Computer Science, vol. 1266, pp. 393–402. Springer, Providence, RI, USA (1997)
28. Rebuge, A., Ferreira, D.R.: Business process analysis in healthcare environments: a methodology based on process mining. Inf. Syst. **37**, 99–116 (2012)
29. Resnik, P.: Using information content to evaluate semantic similarity in a taxonomy. In: Proceedings of IJCAI, pp. 448–453 (1995)

30. Reuter, C., Dadam, P., Rudolph, S., Deiters, W., Trillisch, S.: Guarded process spaces (gps): a navigation system towards creation and dynamic change of helathcare processes from the end-user's perspective. In: Proceedings of Business Process Management Workshops, vol. 2, pp. 237–248 (2012)
31. Socorro, R., Mico, L., Oncina, J.: A fast pivot-based indexing algorithm for metric spaces. Pattern Recogn. Lett. **32**, 1511–1516 (2011)
32. van Elst, L., Aschoff, F.R., Berbardi, A., Maus, H., Schwarz, S.: Weakly-structured workflows for knowledge-intensive tasks: an experimental evaluation. In: Proceedings of 12th IEEE International Workshops on Enabling Technologies (WETICE), Infrastructure for Collaborative Enterprises, pp. 340–345. IEEE Computer Society, Los Alamitos (2003)
33. Weber, B., Reichert, M., Wild, W.: Case-based maintenance for CCBR-based process evolution. In: Roth-Berghofer, T., Goker, M., Altay Guvenir, H. (eds.) Proceedings of European Conference on Case Based Reasoning (ECCBR), 2006. LNAI, vol. 4106, pp. 106–120. Springer, Berlin (2006)
34. Weber, B., Wild, W.: Towards the agile management of business processes. In: Althoff, K.D., Dengel, A., Bergmann, R., Nick, M., Roth-Berghofer, T. (eds.) Professional knowledge management WM 2005. LNCS, vol. 3782, pp. 409–419. Springer, Berlin (2005)
35. Yujian, L., Bo, L.: A normalized Levenshtein distance metric. IEEE Trans. Pattern Anal. Mach. Intell. **29**, 1091–1095 (2007)

Chapter 4
Evaluating a Case-Based Reasoning Architecture for the Intelligent Monitoring of Business Workflows

Stelios Kapetanakis and Miltos Petridis

Abstract CBR-WIMS is a framework that implements the Case-based Reasoning (CBR) process to enable the intelligent monitoring of business workflows. The framework uses the SOA paradigm to store and index a set of business workflow execution event traces. It also allows transparent service interfaces to enterprise system components that orchestrate and monitor business workflows. The CBR component employs similarity measures to retrieve workflow execution cases similar to a given target case. This enables the reuse of associated knowledge about the workflow execution into the new target case. This chapter presents the CBR-WIMS approach and architecture and illustrates its features through its application to two real-life enterprise systems. The chapter examines the portability and robustness of the CBR-WIMS architecture and provides an evaluation of its suitability through an analysis of the experience gained from the two enterprise systems application case studies.

1 Introduction

Modern businesses and organisations use business processes for a variety of purposes. This is due to the constantly evolving operational environments while there is also a need for ensuring formalism and auditability in the business process procedures. They define strictly the internal hierarchical structures as well as setup in a transparent way the relationships among internal and external stakeholders [1]. Business processes offer an agile and flexible way to apply a set business strategies towards a desired goal as well as offer an organisational overview in a leveraged way. Expert managers

S. Kapetanakis (✉) · M. Petridis
School of Computing, Engineering and Mathematics, Moulsecoomb Campus, University of Brighton, Lewes road, Brighton BN2 4GJ, UK
e-mail: s.kapetanakis@brighton.ac.uk

M. Petridis
e-mail: m.petridis@brighton.ac.uk

S. Montani and L. C. Jain (eds.), *Successful Case-based Reasoning Applications-2*,
Studies in Computational Intelligence 494, DOI: 10.1007/978-3-642-38736-4_4,
© Springer-Verlag Berlin Heidelberg 2014

can examine these processes following a bottom-up or top-down approach, focusing on areas of operational significance.

Due to their wide usability and acceptance business processes are being increasingly subject to artificial management and monitoring. Several business process management systems (BPMS) have been implemented following standards that assure the suitable formalism in the representation of business processes. The Business Process Management Notation (BPMN) is a standard of wide acceptance developed to provide a graphical representation of workflow based business processes [2]. The OASIS WS-BPEL is a key execution language, annotating the behaviour of business processes [3]. The XML Process Definition Language (XPDL) provides a standardised format to interchange definitions between workflow vendors [4]. Several other standards like UML Activity Diagrams, BPEL4WS, EPC, and YAWL exist, allowing business process modelling based on different operational perspectives.

Business process standards have been embraced by Service Oriented Architectures (SOAs) allowing the provision of abstracted software web services, loosely connected with requesters. This web service provision through SOA has increased the flexibility and adaptability of BPM systems [5].

With the maturity of adequate representation standards, business processes have ensured their wide acceptance. However, modern business environments are event-driven, and generate a large number of events. As a result a challenge emerges for the effective monitoring and interpretation of a business process instance, based on its generated events. Typically when a business process runs, it generates events that could appear in a general form (name, time-stamp); including though a significant amount of relevant environmental information (detailed name, type, nature, origin, involved workflow stakeholders, further related time-stamps, etc.). The volume of the generated (event-related) data can vary from system to system depending on the system's size, involved operations, the current configuration and the event generation policy followed (trigger an event on every update, on warning, on alert, on severe alert, the list can be expanded further depending on the investigated system).

Event traces generated by a system may be clear and straight-forward when small in number. However, the difficulty in understanding a specific operational statue rises significantly when the numbers of events grow as well as with the underlying system complexity. In many cases the presence of uncertainty, when some of the event information is incomplete or vague can make the job of human auditors harder. This can be a common phenomenon in business workflows that deal with human resources since several events may be not be captured [6] (such as phone calls, unofficial meetings, corridor chats, etc.). In such cases, although the business process can be represented with high precision, its monitoring can be rather difficult, since the system may not be able to identify the relevance of some captured information.

In addressing the above issues, a CBR approach could be elaborated [6] for the intelligent monitoring of workflows based on available past experience. The system could be incorporated within the defined business process and liaise with its associated information drawn from past workflow executions. Based on this information the system could provide relevant diagnosis and explanation [7] on what could be the actual situation faced.

This chapter presents the architecture of CBR-WIMS, a CBR framework developed for the intelligent monitoring of agile business workflows. Section 2 provides a background regarding CBR, work conducted about workflows monitoring as well as the first system used as a case study; Sect. 3 presents the adopted approach for the intelligent workflow monitoring; Sect. 4 presents the architecture of the monitoring framework; Sect. 5 presents an evaluation of the CBR-WIMS based on a new business process monitoring system used as a case study, showing the framework's efficiency and wide reusability of its software components. Finally the conclusion section summarises the work conducted and presents some possible future research paths.

2 CBR and Workflows Monitoring

In order to achieve an effective monitoring of business processes, a variety of approaches can be adopted. When a business process is being monitored, the current series of executed events shows the actual state of the business workflow. This state can be analysed and be compared with existing past knowledge in an attempt for the efficient monitoring of the workflow. The perspective behind such approach is based on the fact that usually similar problems have similar solutions. Case-based reasoning can be regarded as the closest series of techniques to the above perspective following the four R's standard CBR process model [8]. CBR has been proposed as a possible approach to the reuse and adaptation of workflows based on graph representation and the application of structural similarity measures [9]. For the measurement of similarity among business processes Dijkman et al. [10] have investigated algorithms focused on tasks and the control flow relationships among them. Work has also been conducted on the comparison of process models based on behavioural observations within the context of Petri-nets [11]. Similarity measures for structured representations of cases have also been proposed within the context of CBR [12] and have been applied to real life applications requiring reuse of past available knowledge as applied to rich structure-based cases [13, 14].

CBR seems a natural way to allow the effective monitoring of business processes. Existing CBR frameworks provide generic services to assist with the application of CBR on various application areas. An example of this is jColibri [15], an open-source CBR framework aimed at integrated applications that contain models of general domain knowledge. myCBR [16] is another framework providing rapid prototyping of CBR product recommender systems. The above provide effective modelling of applications but do not offer mechanisms for monitoring business processes or environment adaptation according to the investigated process's needs.

A CBR approach towards the intelligent monitoring of business workflows has been proposed and shown to be efficient in monitoring real workflows [6, 7, 17]. This approach can provide workflow monitoring based on similarity measures and a knowledge repository of past cases. Cases consist of several attributes such as event traces, actions taken as well as any available temporal relationships. The characteristics of cases can be represented in terms of a graph used for estimating similarity.

3 CBR-WIMS

CBR-WIMS is a generic architecture and framework developed for the intelligent monitoring provision of business processes [17]. CBR-WIMS provides smooth integration between the CBR knowledge repository and a BPM workflow system. The architecture can deal with the orchestration and choreography of a business process and also provide workflow monitoring. The approach promotes the intelligent management of processes among businesses and organisations. Previous work has shown encouraging results in terms of monitoring been compared to expert managers of specific workflows [6]. The framework can provide a reasoning mechanism for the assistance of workflow managers while dealing with complex situations or situations containing some uncertainty.

Earlier work [17] has shown how the framework's architecture allows smooth integration with a given business process adapting itself to the requirements of the given domain. That work conducted using the CBR-WIMS architecture focused primarily on:

- The adaptation of the architecture on top of a live system forming an integrated case study evaluation.
- Enhancement of the architecture to provide explanation and visualisation facilities required to make the resulting integrated system fully effective.
- Adding elements of inter-connections among business process and reusability in terms of architecture enhancements.

The proposed architecture has now been tested and evaluated with its application on two real-life enterprise systems. The first system is a quality assurance enterprise system used in past work [6, 7, 17]. The second system relates to a Box Tracking System (BTS) which provides temporal tracing of physical items (boxes) within the context of prescribed workflows. The second case study involves advanced temporal complexity deriving from the operational complex context of the workflow. This case study involves certain elements of uncertainty due to the large number of possible errors and alterations that can take place outside of the scope of the workflow monitoring system.

4 Generic Business Process Monitoring

CBR-WIMS is a generic framework and architecture for the intelligent monitoring of Business processes. The motivation behind its design and implementation was to offer a resilient collection of tools that can easily be integrated with an existing business process operational environment. The system should be able to model the rules and limitations of the business process, identify data correlations, actors' roles and activities [17], as well as monitor the execution of the workflow.

4.1 Similarity Measures

In order to be able to compare different execution traces, CBR-WIMS defines the processes in terms of graphs. Similarity measures can be applied afterwards using graph similarity algorithms based on the Maximum Common Sub-graph (MCS) [14]. Sequences of events can be represented as graphs following their temporal execution traces. The relevant information extracted from the available event logs can be represented with the use of the general time theory based on points and intervals [18].

4.2 CBR-WIMS Architecture

The CBR-WIMS system is based on a flexible component-oriented architecture, allowing the collaboration, addition and adaptation of existing components based on the anticipated business process environment. The architecture can be seen at Fig. 1. The full architecture of the system is described in detail in previous published work [17].

CBR-WIMS contains a collection of services that can deal with an imported business process and assist in its effective monitoring. The architecture consists of a set of core software apparatuses that comprise the monitoring backbone and a set of flexible components that can be adapted to a given business process.

The framework is generic. As a result every attempt to use it should start with a process of setting up rules and constraints for a particular formal definition of the business process. The widely accepted standards of BPMN and XPDL are used to import the process business rules and constraints as well as "*sketch*" their operational environment. Based on that, the framework afterwards will be called upon to attempt the process monitoring.

Further to the setup of rules and constraints, CBR-WIMS can adjust its main components in order to be able to deal with the monitored business process. The kernel of the system provides a regulatory context that outlines the overall process. However, the rest of the components are being changed depending on the environmental set. Figure 2 shows the way the framework changes along systems.

For the needs of the BTS system, CBR-WIMS has been subject to a number of modifications. A new Adaptor component has been created in order to create the relevant monitoring parts that can establish monitoring hooks to the BTS workflow

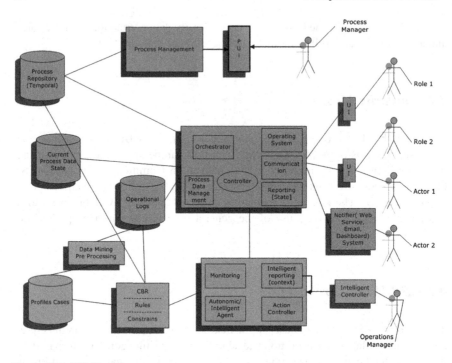

Fig. 1 CBR-WIMS architecture

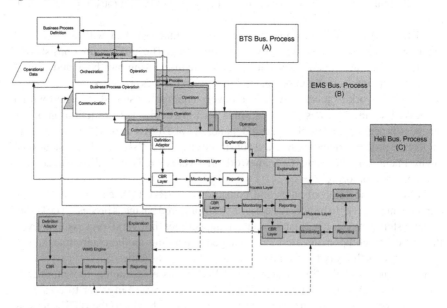

Fig. 2 Architectural layers depending on the investigated business processes

traces. The available logs from the investigated system where parsed using a modified version of the Parser component and were inserted to the WIMS as cases, constituting and populating its case-base. The Adaptor component contains all the communication channels between the investigated system and the monitoring framework ensuring smooth integration of the latter on top of the second's executed traces. The CBR component was not subject to changes, although its case base was populated with different data from what it contained for the previous experiments [6, 7]. Its Similarity measures calculation component was also not affected since its generic methods for the calculation needs were populated explicitly by the Adaptor component.

Figure 2 highlights how the basic architecture of the CBR-WIMS engine can be adapted to different systems (therefore the different colours) based on their native characteristics. The white one refers to the BTS system, the khaki refers to the EMS one [6, 7, 17] and the yellow refers to a working business process tracking system that experiments are currently conducted and will be presented in the future. A point to mention is that in all systems a similar approach is being applied, taking into consideration their workflow representation and the ability to express their business process via rules. In such there can be reuse of the core component of CBR-WIMS and thus its agility is increased across systems.

In order to be able to persist such agility CBR-WIMS does not allow adaptations to core apparatuses when dealing with new business processes. Instead, it requires a semantic ontology to be defined for any investigated system and be populated to the explicit adaptor created for the monitoring purposes. This approach avoids the need for unnecessary adaptations in the system and preserves its integrity and reusability characteristics. A more specific picture that shows the system adaptations for the BTS case study can be seen at Fig. 3. Figure 3 expresses how the core libraries can be encapsulated to more specific (adapted) ones for the needs of BTS system and accompanied by:

(1) Complementary business process components (existing operational systems)
(2) Data from past operation
(3) User Data (profiles, available communications, existing organic hierarchies, etc.)
(4) The actual definition of the business process (rules, BPMN representation, etc.)

In such way the original core of the WIMS can be always expanded to the investigated system's requirements.

Being more explicit in contrast to the system's generic core functions and services, the Results, Reporting and Explanation components are being adapted in order to meet the requirements stemming from the specific monitored business process man-agement system. These components using the framework's visualisation and report compilation libraries, but should also be able to assist the business process experts by responding in their own "*language*" as well as following the explicit characteristics of the system.

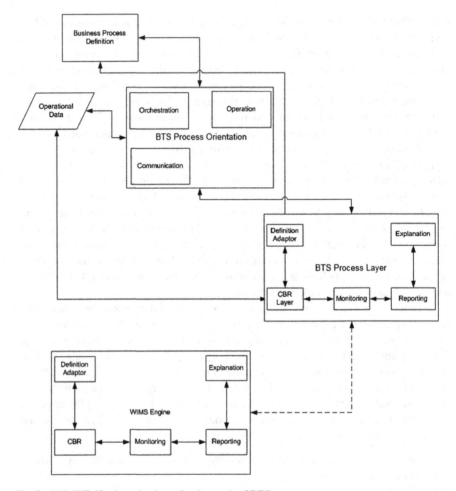

Fig. 3 CBR-WIMS adaptation layer for the needs of BTS system

5 Evaluation

The CBR-WIMS architecture has been first implemented on an educational enterprise system (EMS) [6] that tracked quality assurance transactions and authorisations. That initial evaluation showed the suitability of the CBR-WIMS architecture for the effective monitoring of business workflows. However, that original evaluation did not establish the suitability of the CBR-WIMS architecture as a generic architecture and framework across different types of workflow management systems. Since the original evaluation of the CBR-WIMS framework, a number of the features that ensure its portability and abstraction have been further implemented and enhanced. The evaluation presented in this system concentrates on these specific aspects of the CBR-WIMS architecture and framework. The BTS system was chosen as significantly different from the EMS system. This is both in terms of the nature of the monitored workflows, mainly in terms of the types of events and actors involved and in terms of the different types of enterprise technologies and architectures of the two workflow management systems. For the needs of the new system, several modifications have taken place in terms of the architecture dealing with the system's unique characteristics.

5.1 The BTS Case Study

In order to evaluate the generic characteristic of the proposed architecture, the monitoring on Box Tracking System has been investigated. BTS is an on-line archive management system that deals with the transfer and archiving of boxes between its owned warehouse and its customer premises. Due to the operational nature the system can initiate and host large numbers of transactions that could be spread over time, depending on the customer needs. The number of boxes that could be included in a particular transaction can vary in size (from a rather small number to a rather large one), adding to the given system's complexity. The increased complexity also propagates a higher possibility for errors on either side (company or clients). Box deliveries are being conducted via the BTS owned vehicles. All transactions take place in real-, or near-real- time, and are being monitored via the online system.

System authorised users (managers) are usually monitoring the actual flow of the workflow execution. However, due to the nature of the transactions usually problems are being identified from the customers' side, stating a drawback for the effective monitoring of the business process.

The intelligent monitoring of the executed workflows is built to provide early warning, explanation and context to problems that may emerge at a relatively early stage of developing problems.

5.2 Integration with BTS

In order to evaluate the generic characteristics of the proposed architecture a set of experiments were conducted aiming to verify the adaptability of the architecture to a newly incorporated system. For the evaluation purposes in previous conducted experiments [17], an educational enterprise system was used (EMS) and CBR-WIMS was called to establish a monitoring overview. The system was able to establish similarity measures among workflow execution traces, and present the results to system experts in a visualised way along with elements of textual explanation.

For the needs of these experiments the code structure was reused, although the bare enterprise systems were significantly different from both the implementationand

Table 1 WIMS classification on BTS cases

Cases	CBR-WIMS (%)	Expert's classification
Correctly classified *no clear status* workflows	6–66.7	9
Missed cases	3–33.3	0
False positives	4–19	0
Correctly classified normal workflows	17–81	21

operational perspective. The EMS system [6, 7] is an educational Quality Assurance management system, containing consecutive versions of ASP layers backed up by Access databases. BTS on the other hand is a box management system, comprising of several C-sharp (C#) modules on top of a SQL-server database. CBR-WIMS was able to tackle the integration issues in a rather speedy approach, achieving a smooth, data-integral merge with both systems. Cases of the new system were fed to the case base following the existing proprietary system format. Similarity measures were performed afterwards in a rather trifling time-span. Indicatively the full integration procedure of the system along with the database, that contained more than seventy three thousands of box instances (73,000), lasted approximately eighty (80) work-hours, a rather limited amount of time for operations of that level.

A simple experiment conducted on top of the integrated system included 180 journeys set as the case base. Each journey included a range of parameters indicating certain box states. The MCS [14] was applied to the given cases estimating the similarity measures among them. A sample of 30 journeys was afterwards randomly selected and was redirected for monitoring from both the CBR-WIMS and the system's experts. The experts were able to identify several cases of *"interesting"* behaviour based on the pattern followed within the case. CBR-WIMS was able to identify such patterns and then extrapolate whether a case was of *"interest"* or not based on its closest neighbours' ranking. The system was able afterwards to visualise the similarities between investigated cases and neighbours as well as provide explanation for the provided results. The results provided were able to be shown to system experts without major amendments in WIMS reporting and explanation modules since the flexibility of the architecture allowed their effortless extraction.

System experts were called afterwards to comment on the findings shown by the CBR-WIMS. Since the focus of the experiment was not on the quantitative aspect of monitoring but on the qualitative one, the experts were shown the results and were called to comment on the system's efficiency. It was acknowledged that CBR-WIMS offers a flexible architecture, offering rapid integration to an existing business process due to its high design modularity. Table 1 summarises the results among the experts' monitoring and CBR-WIMS findings. From the results it can be seen that CBR-WIMS can significantly contribute to the experts' monitoring assistance.

6 Conclusions

This chapter has presented the application of CBR-WIMS, an intelligent monitoring platform for agile business workflows. The system was shown to be efficient in monitoring workflow execution cases as well as reusable in terms of its components adaptation. Major advantage of the CBR-WIMS system is its generic architecture which allows a simple step-by-step integration to existing systems using SOA architecture. This chapter has presented its application on two enterprise systems real-life systems coming from different operational disciplines. The system has been proven effective in dynamically assisting human experts in workflow monitoring tasks. Future work will focus more on the operational metrics when applied to different business process environments as well as the in-depth adaptation of the system. The aim is towards integration on current systems. Finally, further work will be concentrated to the investigation of the reusability limits of past experience among different workflows.

References

1. Regev, G., Wegmann, A.: Regulation based linking of strategic goals and business processes. In: Workshop on Goal-Oriented Business Process Modelling, London (2002)
2. Business Process Management Initiative (BPMI): BPMN 2.0: OMG specification. http://www.omg.org/spec/BPMN/2.0 (2011)
3. OASIS: BPEL, the web services business process execution language version 2.0. http://www.oasis-open.org/apps/org/workgroup/wsbpel
4. Workflow Management Coalition (WfMC): XPDL 2.1 complete specification. http://www.wfmc.org/xpdl.html (2008). Accessed 10 Oct 2008
5. Hill, J.B., Sinur, J., Flint, D., Melenovsky, M.J.: Gartner's Position on Business Process Management. Gartner Inc., Stamford (2006)
6. Kapetanakis, S., Petridis, M., Knight, B., Ma, J., Bacon, L.: A case based reasoning approach for the monitoring of business workflows. In: 18th International Conference on Case-Based Reasoning, ICCBR 2010. LNAI, Alessandria, Italy (2010)
7. Kapetanakis, S., Petridis, Ma, J., Bacon, L.: Providing explanations for the intelligent monitoring of business workflows using case-based reasoning. In: Roth-Berghofer, T., Tintarev, N., Leake, D. B., Bahls, D. (eds.) Proceedings of the 5th International Workshop on Explanation-Aware Computing Exact (ECAI 2010), Lisbon, Portugal (2010)
8. Aamodt, A., Plaza, E.: Case-based reasoning; foundational issues, methodological variations, and system approaches. AI Commun. 7(1), 39–59 (1994)
9. Minor, M., Tartakovski, A. and Bergmann, R.: Representation and structure-based similarity assessment for Agile workflows. In: Weber, R.O., Richter, M.M. (eds.) CBR Research and Development, Proceedings of the 7th International Conference on Case-Based Reasoning, ICCBR 2007, Belfast. LNAI, vol. 4626, pp. 224–238. Springer, Berlin (2007)
10. Dijkman, R.M., Dumas, M., Garcia-Banuelos, L.: Graph matching algorithms for business process model similarity search. In Dayal, U., Eder, J. (eds.), Proceedings of the 7th International Conference on Business Process Management. LNCS, vol. 5701, pp. 48–63. Springer, Berlin (2009)
11. van der Aalst, W., Alves de Medeiros, A.K., Weijters, A.: Process equivalence: comparing two process models based on observed behavior. In: Proceedings of BPM 2006. LNCS, vol. 4102, pp. 129–144. Springer, Berlin (2006)

12. Bunke, H., Messmer, B.T.: Similarity measures for structured representations. In: Wess, S., Richter, M., Althoff, K.-D. (eds.) Topics in Case-Based Reasoning. LNCS, vol. 837, pp. 106–118. Springer, Heidelberg (1994)
13. Mileman, T., Knight, B., Petridis, M., Cowell, D., Ewer, J.: Case-based retrieval of 3-D shapes for the design of metal castings. J. Intell. Manuf. Kluwer 13(1), 39–45 (2002)
14. Wolf, M., Petridis, M.: Measuring similarity of software designs using graph matching for CBR. In: Workshop Proceedings of AISEW 2008 at ECAI 2008, Patras, Greece (2008)
15. Recio-García, J.A., Sánchez-Ruiz, A.A., Díaz-Agudo, B., González-Calero, P.A.: jCOLIBRI 1.0 in a nutshell. A software tool for designing CBR systems. In: Proccedings of the 10th UK Workshop on Case Based Reasoning, CMS Press, University of Greenwich (2005)
16. Stahl, A., Roth-Berghofer, Th: Rapid prototyping of CBR applications with the open source tool myCBR. Künstliche Intelligenz 23(1), 34–37 (2009)
17. Kapetanakis, S., Petridis, M., Ma, J., Knight, B.: CBR-WIMS, an intelligent monitoring platform for business processes. In: Petridis, M. (ed.) Proceedings of the 15th UK CBR Workshop, pp. 55–63. CMS press, Cambridge (2010)
18. Ma, J., Knight, B.: A general temporal theory. Comput. J. 37(2), 114–123 (1994)

Chapter 5
The COLIBRI Platform: Tools, Features and Working Examples

Juan A. Recio-García, Belén Díaz-Agudo and Pedro A. González-Calero

Abstract COLIBRI is an open source platform for the development of Case-based reasoning (CBR) systems. It supports the development of different families of specialized CBR systems: from Textual CBR to Knowledge Intensive applications. This chapter provides a functional description of the platform, its capabilities and tools. These features are illustrated with real examples of working systems that have been developed using COLIBRI. This overview should serve to motivate and guide those readers that plan to develop CBR systems and are looking for a tool that eases this task.

1 Introduction

Case-based reasoning (CBR) is a subfield of Artificial Intelligence rooted in the works of Roger Schank in the early 80s, on dynamic memory and the central role that the recall of earlier episodes (cases) and scripts (situation patterns) has in problem solving and learning [1]. In spite of its maturity as a field of research, in 2003 when we started to develop COLIBRI [2], there was no open source tool that would serve as a reference implementation for the common techniques and algorithms developed and refined

Supported by Spanish Ministry of Science & Education (TIN2009-13692-C03-03), Madrid Education Council and UCM (Group 910494), and the Spanish Ministry of Economy and Competitiveness (IPT-2011-1890-430000).

J. A. Recio-García (✉) · B. Díaz-Agudo · P. A. González-Calero
University Complutense of Madrid, Madrid, Spain
e-mail: jareciog@fdi.ucm.es

B. Díaz-Agudo
e-mail: belend@sip.ucm.es

P. A. González-Calero
e-mail: pedro@sip.ucm.es

S. Montani and L. C. Jain (eds.), *Successful Case-based Reasoning Applications-2*,
Studies in Computational Intelligence 494, DOI: 10.1007/978-3-642-38736-4_5,
© Springer-Verlag Berlin Heidelberg 2014

in the CBR community over the years. Our goal was to fill this gap by developing an open source framework that could fulfil several goals: to provide a readily usable implementation of common techniques in CBR useful for academic environments; to foster collaboration among research groups by providing an extensible framework which their prototypes can use and extend; and to promote the repeatability of results achieved by other researchers.

Nowadays, we can conclude that COLIBRI is fulfilling the goal of becoming a reference tool for academia, having, as of this writing, hit the 10,000 download mark with users in 100 different countries. Although the main target of the platform is the research community, a relevant feature of COLIBRI is its support for large-scale commercial applications. It means that our software can be used not only to do rapid prototyping of CBR systems but also to develop applications that will be deployed in real scenarios.

To illustrate the usefulness of the tool some of its applications are outlined here, both in the industry and the academia, that take advantage of the platform.[1] *Aquastress* is a tool to mitigate water stress problems in Europe developed by a consortium of companies and universities under a European Project [3]. Authors point out the out-of-the-box support of COLIBRI for building CBR systems and its compatibility with web applications. Another application developed under a European project that integrates COLIBRI into a web environment is *Ami-Moses*. This is a web system for energy efficiency optimisation in manufacturing companies [4]. Another relevant system using COLIBRI for developing standard CBR systems is *DEEP*, from the US Air Force Research Laboratory, which utilizes past experiences to suggest courses of action for new situations [5].

An important feature of the COLIBRI platform is its support for different families of specialized CBR systems, such as: *Textual*, where the cases are given as text in natural language; *Knowledge-intensive*, where a rich domain model is available and thus the system requires a smaller case base; *Data-intensive*, where cases are the main source of information with no domain knowledge available; or *Distributed*, where different agents collaborate to reach conclusions based on their particular case bases. This way, there are several applications using the extensions provided by COLIBRI to build specialized systems. There is an application for assisting criminal justice [6] that uses the textual capabilities of the platform. Another recommender system developed by an European Project supports users during negotiation processes with recommendations on how to conduct, at the best, a given negotiation [7]. Many of the applications use the knowledge intensive capabilities and point out the facilities to incorporate ontologies and semantic resources in their CBR applications. Some examples are a digital library management system [8], a Web-based catalogue for the electronic presentation of Bulgaria's cultural-historical heritage [9], or a medical application for classifying breast cancer [10].

This chapter describes how COLIBRI supports the development of such applications and provides several examples of working CBR systems that use the features provided

---

[1] The complete list of applications and their references to COLIBRI is available on the project web page http://gaia.fdi.ucm.es/research/colibri/people-users-apps.

by the platform. This overview should serve to motivate and guide those readers that plan to develop CBR systems and are looking for a tool that eases this task.

COLIBRI has been designed following a layered architecture where users can find two different but complementary tools: the jCOLIBRI framework that provides the components required to build programmatically CBR systems; and the COLIBRI STUDIO Development Environment that aids users in the generation of those systems through several graphical tools. The first tool is targeted to developers that prefer to build applications by programming directly, whereas the second one has been created for designer users that prefer high-level composition tools.

COLIBRI offers a collaborative environment where users could share their efforts in implementing CBR applications. It is an open platform where users can contribute with different designs or components that will be reused by other users. COLIBRI proposes a software development process based on the idea of reusing previous designs to aid users in the generation of software systems. In general terms, this process -named the COLIBRI development process- proposes (and promotes) the collaboration among independent entities (research groups, educational institutions, companies) involved in the CBR field. Benefits are manifold: code reuse, automation of system generation, systematic exploration of the CBR design space, validation support, optimization and correct reproducibility are the most significant. Additionally, such capabilities have enormous advantages for educational purposes as students can easily understand and implement complex CBR systems.

Next section introduces the COLIBRI platform and its layered architecture. Then, its development process is presented in Sect. 3 together with the high-level composition tools. Following sections will present the functionality provided by the platform to implement different CBR systems. The basic components are introduced in Sect. 4 and subsequent Sects. 5–9 describe how to build specialized CBR systems. Finally, Sect. 10 contains the related work and Sect. 11 concludes this chapter.

2 The COLIBRI Platform

COLIBRI is an advanced platform that involves several elements. The first building block are the components required to create tangible applications. These components are provided by the jCOLIBRI framework. jCOLIBRI is a mature software framework for developing CBR systems in Java that has evolved over time, built over several years of experience[2] [11].

Once jCOLIBRI was sufficiently mature and has became a reference tool in the CBR community, we have developed the second building block of our platform: an Integrated Development Environment (IDE) that includes graphical tools to aid users in the generation of CBR systems. This environment is called COLIBRI STUDIO[3] and follows a development process based on the idea of reusing previous designs to

---

[2] Download, reference and academic publications can be fount at the web site: www.jcolibri.net.

[3] Available at: www.colibricbrstudio.net.

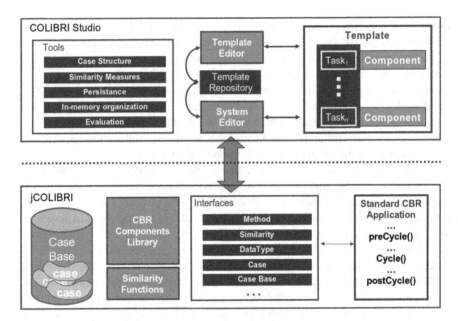

Fig. 1 Two layers architecture of COLIBRI

aid users in the generation of software systems. This development process defines several activities to interchange, publish, retrieve, instantiate and deploy workflows that conceptualize CBR systems. These workflows are called *templates* and comprise CBR system designs which specify behaviour but do not explicitly define functional details.

This way, the COLIBRI platform follows a two-layers architecture where the building blocks of a CBR system are provided by jCOLIBRI and COLIBRI STUDIO enables its composition through the COLIBRI development process. A schema of this organization is shown in Fig. 1. The bottom layer contains the basic components of the jcolibri framework: cases, similarity functions, methods and the interfaces required to implement the core components of a CBR system. The top layer presents an overview of the tools included in COLIBRI Studio.

Following section motivates and presents COLIBRI STUDIO and its associated development process to move later into the description of the capabilities provided by the jCOLIBRI framework.

3 The COLIBRI Development Process

The COLIBRI development process identifies several roles that face the development of CBR systems from different points of view: senior researchers design the behaviour of the application and define the algorithms that will be implemented to create software components that are composed in order to assemble the final system. On the other hand, developers will implement these systems/components. Furthermore, during the last few years there has been an increasing interest in using COLIBRI as a teaching tool, and consequently, our platform also eases this task. These activities conform the COLIBRI development process and are supported by the tools in COLIBRI STUDIO. Next, we describe these activities, tools and user roles:

Template generation.   A template is a workflow-based representation of a CBR system where several tasks are linked together to define the desired behaviour. They should be generated by 'expert' users although other users may create them. This is the first activity in our software development process. Figure 2 shows a simple template in our specialized tool to design templates.

Template publishing.   Templates can be shared with the community. Therefore, there is a second tool that lets users publish a template in the COLIBRI central repository.

Template retrieval and adaptation.   Although the publication of templates is a key element in the platform, the main use case consists of retrieving and adapting a template to generate a new CBR system. Here the actors are not only CBR experts: developers, teachers, students, or inexperienced researchers may perform these activities. Due to their importance, these activities are referred to as the Template

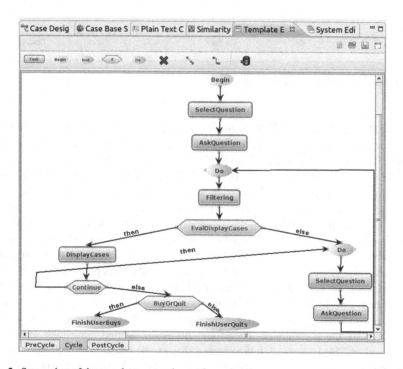

Fig. 2 Screenshot of the template generation tool

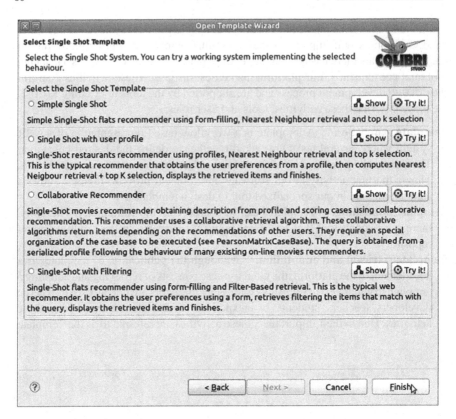

Fig. 3 Screenshot of the template retrieval tool

Base Design (TBD). The TBD begins with the retrieval of a template to be adapted
from the central repository. This retrieval is performed by a recommender system
which proposes the most suitable template depending on the features of the target
CBR system. Figure 3 shows a screenshot of this recommender system. It follows
a "navigation by proposing" approach where templates are suggested to the user.
Next, the adaptation of the retrieved template consists of assigning components
that solve each task. This activity is supported by the tool shown in Fig. 4 where
users can click on a task, obtain a list of compatible components and select the most
suitable one. Then the inputs and outputs of the components can be configured
graphically. The components assigned to tasks are the ones provided by the jCO-
LIBRI framework. For a detailed description of the TBD we point readers to [12].

Component development.   The design of components is closely related to the
advance in CBR research as they implement the different algorithms being discovered by the community. Therefore, this is the second main task of expert
researchers. However, it is not expected that expert researchers will implement
the components. This task will be delegated to users in a 'developer' role. We also
contemplate the role of 'junior researcher' that could design and even implement

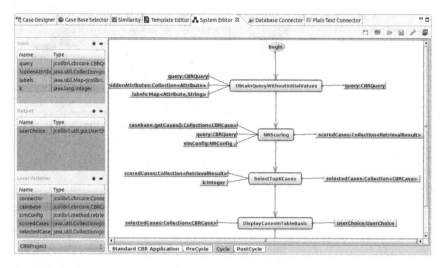

Fig. 4 Screenshot of the template adaptation tool

his own experimental components. Again, these components could be shared with the community by means of the publication tool that uploads it to the COLIBRI repository.

System Evaluation.   As we have mentioned, one of the most relevant benefits of COLIBRI is that it provides an easy to use experimental environment to test new templates and components. Consequently another activity in the development process is the evaluation of the generated systems. It enables the comparison through cross-validation of the performance of different system configurations.

The COLIBRI development process is supported by the tools included in COLI-BRI STUDIO. COLIBRI STUDIO is integrated into the popular Eclipse[4] IDE. This way, it takes advantage of the facilities provided to manage projects and Java source code. It enables the generation of "CBR projects" where the libraries required are automatically configured. It also allows users to compile and run the source code generated by the tools in COLIBRI STUDIO. To begin using COLIBRI STUDIO we provide several wizards and tools that guide the user in the development activity to be performed. For example, the standard wizard lets users configure the following elements of a CBR system: persistence, case structure, similarity measures and in-memory organization of the case base. These tools, partially shown in Fig. 5, are also available in an Eclipse perspective that displays all of them together. In this case Fig. 5 contains the screenshots of the tools used to define the case structure and configure the persistence of the case base.

---

[4] http://www.eclipse.org

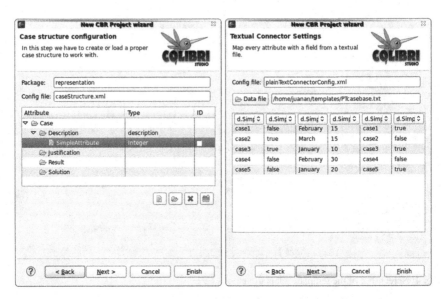

Fig. 5 Screenshots of the wizard tools to configure the case structure (*left*) and persistence (*right*)

The first advantage of our proposal is the reduction of the development cost through the reuse of existing templates and components. This is one of the aspirations of the software industry: that software development advances, at least in part, through a process of reusing components. In this scenario, the problem consists of composing several software components to obtain a system with a certain behaviour. To perform this composition, it is possible to take advantage of previously developed systems. This process has obvious parallels with the CBR cycle consisting on retrieval, reuse, revise and retain. The expected benefits are improvements in programmer productivity and in software quality.

The COLIBRI development process [13] has an additional advantage: the collaboration among users promotes the repeatability of the results achieved by other researchers. Nowadays, the reliability of the experimental results must be backed up by the reproducibility of the experiments. This feature ensures the advance in a research area because further experiments can be easily run by extending the existing ones. It is a development process that promotes the reproducibility of experiments for the CBR realm.

An extended description of the COLIBRI development process and COLIBRI STUDIO can be found in [14]. We point readers to that paper for details. Now that we have outlined how to build CBR systems with the COLIBRI platform we detail what kind of systems can be implemented. The implementation is based on the components available in the jCOLIBRI framework, therefore its capabilities are explained next.

4 A Functional Description of the jCOLIBRI **Framework**

Addressing the task of developing a CBR system raises many design questions: How are cases represented? Where is the case base stored and how are the cases loaded? How should algorithms access the information inside cases? How is background knowledge included? and so on. The best way to solve these issues is to turn to the expertise obtained from previous developments. Thus, COLIBRI defines how to design CBR systems and their composing elements. This definition of the structure of a CBR system is a key element in the platform as it enables the compatibility and reuse of components and templates created by independent sources. The main elements of this structural design are:

1. Organization into persistence, core and presentation layers. It follows the structure used by J2EE[5] where persistence, business logic and presentation are independent layers.
2. Definition of a clear and common structure for the basic objects found in a CBR application: cases, queries, connectors, similarity metrics, case-base organization, etc. For example, in COLIBRI a query is the description of a problem, and a case is an extension of a query that adds the solution to that problem, a justification of that solution and, probably, a record of the result of applying the case in the real world. This organization was decided after a revision of several works found in the CBR literature [15, 16].
3. Run-time structuring of the CBR systems: jCOLIBRI organizes the behaviour of the CBR systems into: precycle, where required resources (mostly cases) are loaded; cycle, which performs the 4 R's tasks that structure the reasoning (retrieve, reuse, revise and retain [17]); and postcycle, which releases resources.

This structural design of the CBR systems requires a reference implementation that enables users to create tangible applications. This implementation is provided by the jCOLIBRI framework. It includes the components required to build CBR systems both programmatically or through the COLIBRI STUDIO composition tools:

- Connectors. Cases can be stored using different media. Common media are databases or plain text. However, there may be any other potential source of cases that can be exploited by the framework, such as ontologies, which are accessed through Description Logic reasoners like RACER [18]. Therefore, jCOLIBRI defines a family of components called *connectors* to load the cases from different media into the in-memory organization: the *case base*. This division into two layers enables an efficient management of cases, an issue that becomes more relevant as the size of the case base grows. This design based on connectors and in-memory storage of the case base is shown in Fig. 6.
- Cases. jCOLIBRI represents the cases using *Java Beans*. Using Java Beans in jCOLIBRI, developers can design their cases as normal Java classes, choosing the most natural design. This representation simplifies programming and debugging

---

[5] *Java 2 Enterprise Edition.*

Fig. 6 Persistance organization in jCOLIBRI. It includes
connectors for databases,
textual files, ontologies and
Weka's ARFF format

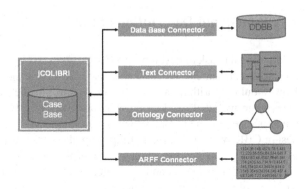

CBR applications, and configuration files became simpler because most of the metadata of the cases can be extracted using the introspection capabilities of the Java platform. Java Beans also offer automatically generated user interfaces that allow the modification of their attributes, and direct persistence into databases and XML files. It is important to note that many Java web applications use Java Beans as a base technology, so the development of web interfaces is very straightforward. Moreover, Hibernate[6] —the library used to develop the database connector in jCOLIBRI—uses Java Beans to store information in a database. Java Beans and Hibernate are core technologies in the *J2EE* platform. By using these technologies in jCOLIBRI, we guarantee the possibility of integrating CBR applications developed using this framework into large scale commercial systems. [19] provides an extended description of case representation in the framework.

- Case base. There are several components to organize the case base once cases have been loaded from the persistance media. These organizations use different data structures such as: linear lists, trees, hash maps, etc.
- Retrieval methods. The most important retrieval method is Nearest Neighbour scoring. It uses global similarity functions to compare compound attributes and local similarity functions in order to compare simple attributes. Although this method is very popular, there are other methods that are also included in the framework. For example, we implement Expert Clerk Median scoring from [20] and a filtering method that selects cases according to boolean conditions on the attributes. Both methods belong to the recommenders field and will be explained in Sect. 6. In the textual CBR field, we also find specialized methods using several Information Retrieval or Information Extraction libraries (Apache Lucene, GATE, ...) that will be detailed in Sect. 5. Regarding knowledge intensive CBR, we enable retrieval from ontologies (Sect. 7). Data intensive retrieval is addressed by means of clustering algorithms and a connector for the Weka ARFF format (Sect. 8). And, finally, jCOLIBRI provides the infrastructure required to retrieve cases in a distributed architecture of agents (Sect. 9).

---

[6] http://www.hibernate.org

Once cases are retrieved, the best ones are selected. The framework includes methods to select the best scored cases but also provides diversity metrics. Most of these methods belong to the recommender domain, therefore they are detailed in Sect. 6.

- Reuse and revision methods. These two stages are coupled to the specific domain of the application, so jCOLIBRI only includes simple methods to copy the solution from the case to the query, to copy only some attributes, or to compute direct proportions between the description and solution attributes. There are also specialized methods to adapt cases using ontologies that will be explained in Sect. 7.

- Retain. These methods are in charge of adding new cases to the case base. There are strategies to reflect the changes to the persistence media or just modify the in-memory storage.

- Evaluation tools measure the performance of a CBR application. jCOLIBRI includes the following cross-validation strategies: *Hold Out, Leave One Out* and *N-Fold*. A detailed explanation of these evaluation tools is presented by [19].

- Maintenance. These methods allow developers to reduce the size of the case base.[7] Components provided are: BBNR (Blame-based noise reduction) and CRR (Conservative Redundancy Removal) [21], RENN (Repeated Edited Nearest Neighbour) [22], RC (Relative Cover) [23], or ICF (Iterative Case Filtering) [24].

- Visualization. These methods represent graphically the similarity between cases.[8] This tool serves to debug the fitness of the similarity measure and is shown in Fig. 7 It assigns a color to each type of solution and lays out the cases according to their similarity.

Summarizing, jCOLIBRI offers 5 different retrieval strategies with 7 selection methods and provides more than 30 similarity metrics. It provides around 20 adaptation and maintenance methods plus several extra tools like system evaluation or the visualization of the case base.

Now that we have described the structure and functionality provided by the framework, the following sections detail how to build specialized CBR systems. The components required to build such systems are delivered as packages named *extensions*, that offer additional services and behaviour beyond basic CBR processes. The most relevant extensions are included in the main distribution but others can be obtained separately at the web site.[9] This web site also provides *contributions*, i.e. extensions developed by third-party research teams. Extensions include several examples to let developers understand the services provided. Then, the API documentation details how to use and integrate each component into an existing development. Following sections describe the most relevant extensions provided by jCOLIBRI to offer a complete overview of the framework's capabilities.

---

[7] Maintenance components were contributed to the library by Lisa Cummins and Derek Bridge from the University College of Cork, Ireland.

[8] The visualization facility was contributed by Josep Lluis Arcos (Artificial Intelligence Research Institute, Spain).

[9] http://gaia.fdi.ucm.es/research/colibri/jcolibri/contributions

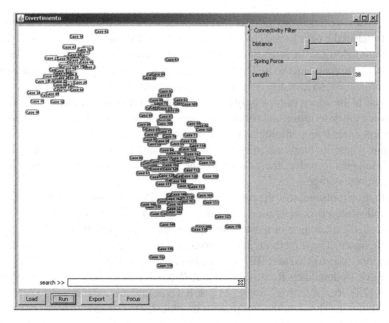

Fig. 7 Screenshot of the evaluation tool [19]

5 Textual CBR Applications

Textual CBR (TCBR) is a subfield of CBR concerned with research and implementation on case-based reasoners where some or all of the knowledge sources are available in textual format. It aims at using these textual knowledge sources in an automated or semi-automated way to support problem-solving through case comparison [25].

Although it is difficult to establish a common functionality for TCBR systems, several researchers have attempted to define the different requirements for TCBR [25, 26]: how to assess similarity between textually represented cases, how to map from texts to structured case representations, how to adapt textual cases, and how to automatically generate representations for TCBR.

The textual CBR extension of jCOLIBRI provides methods to address some of these requirements using different strategies. These sets of methods are organized according to:

1. A library of CBR methods solving the tasks defined by the Lenz layered model [27].[10] The goal of these layers is to extract the information from the text into a structured representation that can be managed by a typical CBR application. This way, the CBR application can perform a retrieval algorithm based on the similarity of the extracted features and use them to adapt the text to the query.

---

[10] This extension was developed in collaboration with Nirmalie Wiratunga from The Robert Gordon University, U.K.

These layers are typical processes in Information Extraction (IE) systems: stemming, stop words removal, Part-of-Speech tagging, text extraction, etc. jCOLIBRI includes several implementations of these theoretical layers. A first group of methods uses the Maximum Entropy algorithms provided in the OpenNLP package.[11] The second implementation uses the popular GATE library for text processing.[12] Reference [28] exemplifies these methods to annotate web pages and perform semantic retrieval. This way, we provide a structured representation of cases that can be managed by standard similarity matching techniques from CBR. The main disadvantage of these methods is that they can only be used where texts are mapped to cases with a fixed structure (cases always have the same attributes). An example is the restaurant adviser we use in [29]. Another drawback is that it requires the definition of IE rules for each concrete domain, increasing the development cost. We point readers to that paper [29] for further details about the semantic TCBR methods in jCOLIBRI.

2. A natural language interaction module that facilitates querying CBR systems by using written natural language. IE techniques and Description Logics (DLs) reasoning [30] play a fundamental role in this interaction module, where it analyses a textual query from the user to generate a structured representation using the relevant information from the text. The *understanding* process, from the textual query to the structured query, follows the steps of *IE*, *Synonyms*, *Reasoning with Ontologies* and *User Confirmation*, where the user validates or corrects the extracted information. The details and an evaluation of this module are presented by [31]. Figure 8 shows a screenshot of the application described in that paper.

3. There is another group of textual CBR methods that can be used when cases are texts without a fixed structure [32]. These methods are based on Information Retrieval (IR) conjointly to clustering techniques where reuse and retrieval are performed in an interleaved way. This third approach is clearly different from

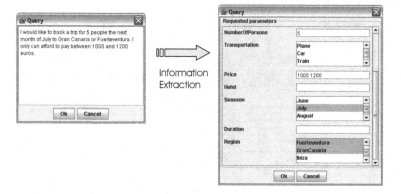

Fig. 8 Natural language interaction module

---

[11] http://opennlp.sourceforge.net
[12] http://gate.ac.uk

the two previous approaches, which are based on IE techniques to "capture the meaning", i.e, to engineer structured case representations. This set of textual methods belongs to the statistical approach that has given such good results in the IR field. jCOLIBRI methods are based on the Apache Lucene search engine [33]. Lucene uses a combination of the Vector Space Model (VSM) of IR and the Boolean model to determine how relevant a given document is to a user's query. The main advantages of this search method are its positive results and its applicability to non-structured texts. The big drawback is the lack of knowledge regarding the semantics of the texts.

5.1 A TCBR Application Built with COLIBRI

Based on the new methods included in jCOLIBRI, in [32] we present a complete Textual CBR application to deal with the automatic generation of failure reports. This system retrieves textual reports and guides the user to adapt them to the current situation. The absence of experts leads us to use IR+Clusters instead of Information Extraction. Figure 9 contains an screenshot of this application—named Challenger—that illustrates the steps followed by the user to adapt a text.

Although statistical IR methods give good retrieval results they do not provide any kind of explanation about the documents returned. One way for solving this problem is to cluster the retrieval results into groups of documents with common information. Usually, clustering algorithms like hierarchical clustering or K-means

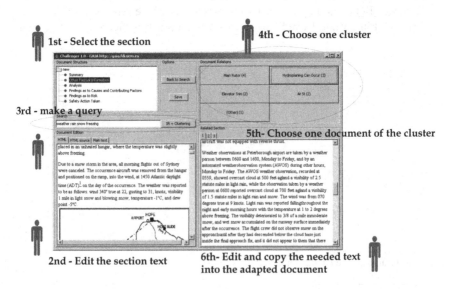

Fig. 9 Screenshot of the Challenger application

[34] group the documents but they don't provide a comprehensive description of the resulting clusters. Lingo [35] is a clustering algorithm implemented in the Carrot2 framework[13] that allows the grouping of search results but also gives the user a brief textual description of each cluster. Lingo is based on the Vector Space Model, Latent Semantic Indexing and Singular Value Decomposition to ensure that there are human readable descriptions of the clusters and then to assign documents to each one. jCO-LIBRI provides wrapper methods to hide the complexity of the algorithm and allows a simple way for managing Carrot2.

This labeled-clustering algorithm can be applied to TCBR in the retrieval step to make it easier to choose the most similar document. However, in [32] we present an alternative approach that uses the labels of the clusters to guide the adaptation of the texts. The adaptation of texts is a knowledge intensive task, especially for non-structured and technical texts. So, we propose a software solution whereby a user is provided with interactive tools to assist with adapting a retrieved report. This way, the user is in charge of the adaptation although the system finds the knowledge required to perform this task by looking for similar pieces of texts.

This method is described as a transformational reuse method where one copy of the case retrieved is used as the source case. Then the user makes changes using a text editor: deleting text components (words, phrases, paragraphs, or sections); writing new text components; or substituting text components using other similar text components. Our method assists the user in the substitution step, as it uses the clusters to show which are the text components that are related to the piece of text that the user is currently adapting. To clarify, we are proposing a simple way of adaptation where the user makes the changes from the choices we provide. These choices are based on the IR and clustering techniques previously explained so that, in a way, the adaptation process is an extension of retrieval, where the system retrieves good candidates to do substitutions.

This approach would be quite similar to the one presented in [36], although our supervised approach allows us to deal with texts with different structure.

For interested readers, [19] presents an extended tutorial on textual CBR with jCOLIBRI.

6 Recommender Systems

CBR has played a key role in the development of several classes of recommender systems [37, 38]. The jCOLIBRI extension for building recommender systems is based in part on the conceptual framework described in the paper by [37][14] that reviews different recommendation approaches. The framework distinguishes between collaborative and case-based, reactive and proactive, single-shot and conversational,

---

[13] http://project.carrot2.org/

[14] The recommenders extension has been developed in collaboration with Derek Bridge from University College Cork, Ireland.

and asking and proposing. Within this framework, the authors review a selection of papers from the case-based recommender systems literature, covering the development of these systems over the last ten years (we could cite [20, 39–42] as illustrative examples). Based on this revision, jCOLIBRI includes the following set of methods:

- *Methods for retrieval.* Different approaches to obtain and rank items to be presented to the user: typical similarity-based retrieval [39], filter-based retrieval based on boolean conditions, ExpertClerk's median method [20] and collaborative retrieval method based on users' ratings to predict candidate items [43, 44].
- *Methods for case selection.* Approaches to select one or more items from the set of items returned by the retrieval method: (1) Select all or k cases from the retrieval set. (2) The Compromise-driven selection [45] method chooses cases according to a number of attributes compatible with the user's preferences. (3) Finally, the Greedy Selection [46] method considers both similarity and diversity (inverse of similarity).
- *Methods for navigation by asking.* Navigation by asking is a conversational strategy where the user is repeatedly asked about attributes until the query is good enough to retrieve relevant items. There are different methods based on heuristics to select the next best attribute to ask about. For example, Information Gain, which returns the attribute with the greatest information gain in a set of items [40, 47], and Similarity Influence Measure [40], which selects the attribute that has the highest influence on Nearest Neighbor similarity.
- *Methods for navigation by proposing.* The navigation by proposing strategy asks the user to select and critique one of the items recommended. The selected item is modified according to the critique and produces a new query. There are different strategies to modify the query as enumerated by [48]. The More Like This (MLT) strategy replaces the current query with the description of the selected case. Partial More Like This (pMLT) strategy partially replaces the current query with the description of the selected case but it only transfers a feature value from the selected case if none of the rejected cases have the same feature value. Another option is to use MLT but weighting the attributes (Weighted More Like This, wMLT). Less Like This (LLT) is a simple one: if all the rejected cases have the same feature-value combination, which is different from the preferred case, then this combination can be added as a negative condition. Finally, More + Less Like This (M+LLT) combines both More Like This and Less Like This.

There are other methods to display item lists, make critiques, display attribute questions, manage user profiles, etc. Details are provided by [49] and the jCOLIBRI code examples.

6.1 A Recommender Application Built with COLIBRI

To illustrate the capabilities of the recommender extension, we will present the HappyMovie application. HappyMovie is a recommender system that provides a recommendation for a group of people that wish to go to the cinema together. Figure 10 includes a screenshot of the system. It is a Facebook application that exploits social knowledge to reproduce more concisely the real argumentations made by real groups of users. Its generic architecture –named ARISE (Architecture for Recommendations Including Social Elements)—is depicted in Fig. 11.

First, it uses a personality test in order to obtain the different roles that people play when interacting in a decision making process. Additionally, the tie-strenght (or trust) between users is also included in the recommendation method by extracting specific information from their profiles in the social network.

The satisfaction database represents the "memory" of the system for future recommendations. This module stores all the recommendations that have been made for every user and every group. Having recommendations with memory allows our system to avoid previous recommendations so that it does not repeat itself and allows also to ensure a certain degree of fairness. If one member accepts a proposal that she is not interested in, next time she will have some kind of preference, so that in the long run all the members of the group are equally satisfied. This module is actually a CBR system where cases are the previous recommendations.

Fig. 10 Screenshot of the HappyMovie system

Fig. 11 Generic architecture of the HappyMovie system

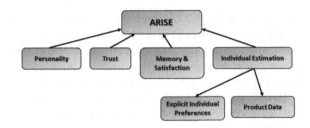

The recommendation strategies in HappyMovie predict the rating that each user would assign to every item in the catalogue and then these estimated ratings are combined to obtain a global prediction for the group. Then, the movie with the highest prediction is proposed. Therefore, a basic building block of this application is the module in charge of computing individual predictions. These individual predictions are obtained by applying the retrieval methods in jCOLIBRI that take into account the personal preferences of the user and the available movies.

Additional details of this system can be found in [50].

7 Knowledge Intensive CBR

Our research group[15] has been working for more than a decade on knowledge intensive CBR (KI-CBR) using ontologies [51–54]. Commonly, KI-CBR is appropriate when developers do not have enough experiences available but there is a considerable amount of knowledge of the domain. [19] provides further details about the implementation of KI-CBR applications in jCOLIBRI.

We state that the formalization of ontologies is useful for the CBR community regarding different purposes, and therefore, jCOLIBRI supports the following services:

1. Persistence: jCOLIBRI provides a connector that loads cases represented as concepts or individuals in an ontology (see Fig. 6). This connector delegates to our library for managing ontologies named *OntoBridge*.
2. Definition of the case structure through elements in ontologies. This extension provides a specialized data type used to represent attributes of the case structure that point to elements in an ontology. For example, an attribute *city* used in the representation of a case can be linked to the concept *City* in an ontology. This way, cases following that structure will store the values *Madrid, London, N.Y., Tokyo* in the attribute *city* because they are the individuals belonging to the concept *City* in the ontology.

   This approach that links case attributes to elements from an ontology can be used either if the cases are embedded as individuals in the ontology itself, or if the

---

[15] http://gaia.fdi.ucm.es/

cases are stored in a different persistence medium, such as a database, but some attributes contain values from the ontology.

3. Retrieval and similarity. There are different strategies to compute the local similarity based on ontologies [55–57]. Following the previous example, a more elaborate ontology will classify cities according to continents. Therefore the concept *city* will be specialized by the subconcepts *EuropeanCity, AmericanCity, AsianCity,* ... the individuals being organized consequently. This way we can use this hierarchy to compute similarity according to the distance between individuals. In this case, the similarity between *Madrid* and *London* will be higher than *Madrid* and *Tokyo* because *Madrid* and *London* belong to the same subconcept. This approach to compute similarity based on the distances—named *concept-based similarity*—can be performed in different ways [55], and jCOLIBRI provides the implementation of these similarity metrics to developers.

4. Adaptation. The usage of ontologies is especially interesting for case adaptation [51], as they facilitate the definition of domain-independent, but knowledge-rich adaptation methods. For example, imagine that we need to modify the *city* attribute of a retrieved case because the current value *Madrid* is not compatible with the restrictions of the query. According to the ontology, the best substitute is *London* because it is also a *EuropeanCity*. Here again we use the structure of the ontology to adapt the case. Because this schema is based on distances within the ontology, jCOLIBRI offers several domain-independent methods to perform this kind of adaptation. These methods only need to be configured with the ontology (which contains the domain knowledge).

   Following subsection presents two adaptation methods included in jCOLIBRI. See [51, 58, 59] for a detailed description of different adaptation proposals.

5. Learning [60]. As ontologies are used as a persistence media, ontologies can be reused to store the experiences learnt. This is performed by means of the connector able to manage ontologies.

7.1 A KI-CBR Applicacion Built with COLIBRI

A KI-CBR system using the adaptation methods included in jCOLIBRI is described in [59]. This paper presents a folk tale generation system that analyzes the role of reuse in CBR systems in originality driven tasks, where a new solution has not only to be corrected but noticeably different from the ones known in the case base. Each case is a story plot that, is formalized by its actions, and each action by its properties, like the participant characters and their roles (Donor, Hero, FalseHero, Prisoner, Villain), the place where the action takes place (City, Country, Dwelling), the involved objects, attributive elements or accessories (a ring, a horse). Cases must keep a semantic coherence that ensures the dependencies between actions, characters and objects.

For example, Release-from-Captivity and Kidnapping, or Resurrection and Dead functions. Therefore, ontologies are the best approach to represent this kind of cases. Each case is composed of a great number of interrelated individuals, i.e instances of concepts, from the ontology. A visual representation of the tales ontology is presented in Fig. 12.

To generate new tales, the system applies two broadly different Reuse techniques, one based on transforming an existing solution into a new solution and another based on generating or constructing a new solution.

Transformational Reuse—or Transformational Adaptation (TA)—is the most widely used approach to case reuse. Typically, a new case is solved by retrieving the most similar case and copying the solution (although some techniques may use solutions from multiple cases); then a transformational process using domain knowledge and/or case-derived knowledge modifies that copy (which we consider a form of search) until a final solution adequate to the current problem is found. Basically, a node in the "working case" is substituted by finding another related node in a taxonomic hierarchy—e.g. a sword is a type of weapon in the folk tale generation domain, and may be substituted by another weapon like a crossbow. Moreover, Transformational Reuse is able to modify more than a single node: deep substitution allows to modify a whole subgraph in the solution—e.g. when substituting a character like the evil wolf for an evil wizard, then the constituent aspects of the characters (role, sex, dwelling, physical appearance) are also substituted.

Generative or Constructive Reuse builds a new solution for the new case while using the case base as a resource for guiding the constructive process. Constructive Adaptation (CA)[61] is based on a heuristic search-based process where the heuristic function guiding search is derived from a similarity measure between the query and the case base. This method takes a problem case and translates it into an initial state in the state space; i.e. transform a case representation into a state representation. Then, a heuristic search process expands a search tree where each node represents a partial solution, until a final state (with a complete and valid solution) is found. Notice that final but non-valid states can be reached, but this simply means the search process will backtrack to expand other pending states. This process is guided by a heuristic based on comparing the similarity from states (represented in the state space) to cases (represented in the space for cases). The nodes with higher similarity are expanded first during the search process. The result is that CA adds one node to a partial solution as it moves from one state to the next; that is to say, it builds a solution by piecemeal copies of nodes from similar cases. Notice that there is neither retrieval nor "single case adaptation" here since the component nodes are incrementally copied from multiple cases in the case base, depending only on the similarity measure that works on the whole case base.

Now that we have explained the main methods in jCOLIBRI for KI-CBR we can move on to an opposite family of CBR systems where many cases are used instead of using few but rich ones.

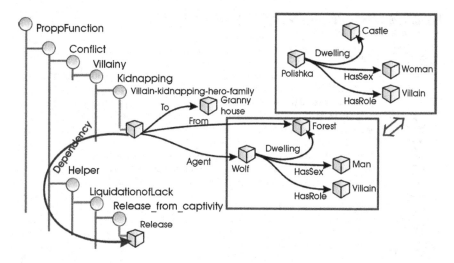

Fig. 12 Semantic dependencies in the folk tales ontology

8 Data Intensive CBR

One of the problems to solve when dealing with real world problems is the efficient retrieval of cases when the case base is huge and/or it contains uncertainty and
partial knowledge. There are many examples of domains and applications where a
huge amount of data arises; for example, image processing, personal records, recommender systems, textual sources, and many others. Many authors have focused on
proposing case memory organizations to improve retrieval performance. For example, there are different proposals to manage huge case memories organized in clusters
such as the ones by [62, 63]. However none of the existing tools has incorporated
capabilities to efficiently manage large case bases. jCOLIBRI provides an extension
called *Thunder* to address this issue. Thunder[16] allows CBR experts to manage case
memories organized in clusters and incorporates a case memory organization model
based on Self-Organizing Maps (SOM) [64] as the clustering technique. This extension includes a graphical interface to test the components provided as shown in
Fig. 13.

Clustering is implemented by grouping cases according to their similarities and
representing each one of these groups by prototypes. Thus, the retrieve phase carries
out a selective retrieval focused on using only the subset of cases potentially similar
to the new case to solve. The new case retrieval procedure consists of (1) selecting
the most suitable group of cases by comparing the input case with the prototypes
and, (2) comparing the new input case with the cases from the selected clusters.
The benefits of such an approach are both the reduction of computational time and

[16] Thunder was developed in collaboration with Albert Fornells, from Universitat Ramon Llull,
Spain. Available at: http://gaia.fdi.ucm.es/research/colibri/jcolibri/contributions.

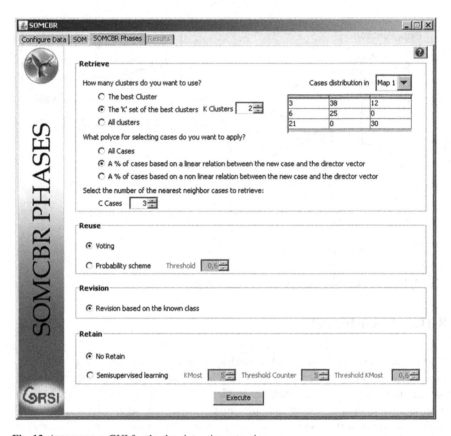

Fig. 13 jCOLIBRI test GUI for the data intensive extension

improved robustness with uncertain data. Nevertheless, some open issues remain such as to what extent the accuracy rate is degraded due to the cluster-based retrieval, and furthermore how many clusters and cases should be used according to given requirements of computational time and accuracy degradation.

To support a uniform way for loading these large case bases, the Thunder extension provides a connector compatible with the ARFF format. This format is a standard defined by the popular Weka [17] toolkit for data mining and machine learning [65].

8.1 A DI-CBR Application Build with COLIBRI

To illustrate the DI-CBR methods in COLIBRI we present an application, described in [66], that serves to classify automatically documents from electronic journals into

---

[17] http://www.cs.waikato.ac.nz/ml/weka/

different categories: laws, history, medicine, ... This system is used by librarians to assign proper bibliographic categories when a new text is included into the catalogue. The application is also a Textual CBR system as it manages textual data. In this case, a statistical similarity function is used to compare documents. The system uses a corpus of 1.500 documents belonging to 20 different categories. Documents are processed to remove stop words and extract the stem. Then the TF*IDF filter is applied to select the most relevant 1.000 terms in the corpus. This corpus is a clear example of a huge case base with uncertain and incomplete cases.

This CBR system uses a majority-voting approach that assigns the most repeated category of the k nearest neighbors. The performance tests run over the case base shown the improvement in the efficiency of the system when using a clustered memory of cases. A key parameter of this system is the number of clusters being considered. For example, when comparing the query with the prototype of the 7 most similar clusters, the precision only decreases from 10%, meanwhile the number of case being compared is half the size of the case base (47%). This important reduction in the cases involved in the retrieval process implies a significant improvement in the efficiency of the CBR system.

Figure 14 illustrates this process with a reduced number of cases. Each color represents a different category. As we can observe, the clustering algorithm trends to group similar cases that should belong to the same category. Once each cluster has been identified, the query is compared with the corresponding prototype.

Next we describe the last extension of jCOLIBRI: an infrastructure for developing distributed CBR systems.

Fig. 14 Visualization of a clustered case-base

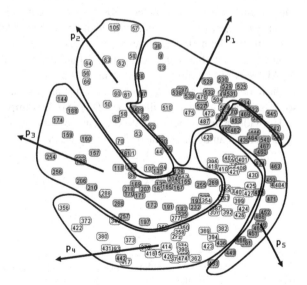

9 Distributed CBR

Research efforts in the area of *distributed CBR* concentrate on the distribution of resources within CBR architectures and study how it is beneficial in a variety of application contexts. In contrast to single-agent CBR systems, multi-agent systems distribute the case base itself and/or some aspects of the reasoning among several agents. [67] categorized the research efforts in the area of distributed CBR using two criteria: (1) how knowledge is organised/managed within the system (i.e. single vs. multiple case bases), and (2) how knowledge is processed by the system (i.e. single vs. multiple processing agents).

Much of the work in distributed CBR assumes multi-case base architectures involving multiple processing agents that differ in their problem solving experiences [68]. The "ensemble effect" [69] shows that a collection of agents with uncorrelated case bases improves the accuracy of any individual. Multiple sources of experience exist when several CBR agents need to coordinate, collaborate, and communicate. Within this purpose, jCOLIBRI provides two extensions to design deliberative and distributed multiagent CBR systems where the case base itself and/or some aspects of the reasoning process are distributed among several agents. A deliberative system can predict the future state that will result from the application of intentional actions. These predicted future states can be used to choose between different possible courses of actions in an attempt to achieve system goals [70]. Our work focuses on distributed *retrieval* processes working in a network of collaborating CBR systems.

The basic extension to support distributed CBR applications in jCOLIBRI is called *ALADIN* (Abstract Layer for Distributed Infrastructures). This layer defines the main components of every distributed CBR system: agents, directory, messages, etc. and could be implemented by using different alternatives: JADE[18], sockets, shared memory, ... It was defined after reviewing the existing literature on distributed CBR and is mostly compatible with IEEE FIPA[19] standards for multiagent systems. Because ALADIN is only composed of interfaces that define the behaviour of the system, we have developed an implementation of this abstract layer using standard network sockets. This extension is called *SALADIN* (Sockets implementation of ALADIN[20]) and provides a fully functional multi-agent environment for building distributed CBR systems that can be particularized in many ways.

---

[18] JADE is a framework for building multi-agent systems following the FIPA specifications. It is available at: http://jade.tilab.com/.

[19] Foundation for Intelligent Physical Agents. http://www.fipa.org/.

[20] ALADIN and SALADIN extensions are available at: http://gaia.fdi.ucm.es/research/colibri/jcolibri/contributions.

9.1 A Distributed CBR System Built with COLIBRI

In this section we describe a distributed CBR application in the domain of music recommendation, a classical example of successful recommender applications where there are many users interested on finding and discovering new music that would fulfill their preferences. Moreover, the users of this kind of applications tend to interchange recommendations with other users that have similar preferences. These relationships conform social networks that reflect the similarity in the preferences of the users and allow them to discover new items, and the confidence in the recommendation. This way, the system can measure the confidence between two users depending on their corresponding distance in the social network.

As the case base has a catalogue of songs, each user may have a part of this catalogue in its internal list of rated items. Every user interacts with its corresponding recommender agent. When a recommender agent receives a query from the user, it forwards the query to the other agents in the system it is connected to. Agents are organized according to a social network that ideally reflects the similarity and confidence between users. Then, these agents use their rated items to recommender songs that fulfil the preferences of the query. This organization according the social network reports a higher performance than the typical all-connected configurations. Both organizations are illustrated in Fig. 15.

For further details we refer the interested reader to the paper by [71].

10 Related Work

COLIBRI is nowadays the most popular CBR platform due to its broad scope, number of applications, users and contributors. There are other related CBR tools in the literature, with myCBR [72] being one of the most closely related to jCOLIBRI. myCBR

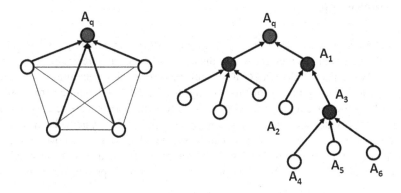

Fig. 15 Agents organization in a distributed CBR system

is also an open-source tool for developing CBR systems, although there are some important differences in its scope and architecture. myCBR is intended primarily for prototyping applications that focus on the similarity-based retrieval step while CO-LIBRI includes components for supporting the whole CBR cycle, including retrieval, reuse, revision and adaptation. To a certain extent myCBR and COLIBRI can be used in collaboration, using myCBR to define the case structure and similarity measures through their Protégé-based interface, and having COLIBRI import those definitions from the XML files generated by myCBR, through a number of wrapper methods that were developed in collaboration by both project teams.[21]

The Indiana University CBR Framework (IUCBRF[22]) was conceived with a similar goal of providing an open-source solution for academic CBR projects but never achieved a mature enough development state and has not been actively maintained in the last years [73].

Although in a different domain jCOLIBRI is also related to Weka, a collection of machine-learning algorithms for data-mining tasks. Weka originally served as an inspiration for COLIBRI which was intended to play a similar role in the CBR community, as an open-source reference tool in academia, like the one played by Weka in the data mining community. COLIBRI is also influenced by Weka in the way it is designed to facilitate the construction of different configurations of a CBR system which can be compared along different dimensions. The two tools also share the idea of including both a collection of reusable methods, plus a software composition tool that allows to assemble a running system without writing a line of code. They even share some common methods, since Weka also includes an implementation of the Nearest Neighbours algorithm for instance-based classification. Nevertheless, and in addition to the obvious differences coming from covering different application domains, a key difference between the tools is that COLIBRI is designed for supporting not only classification but also problem-solving tasks, and can deal with complex object-based data and not only simple attribute-value pairs like Weka.

11 Conclusions

In this chapter, we have described the COLIBRI platform for building CBR systems. The mail goal of our research is to cover the need for open software engineering tools for CBR. We can conclude that COLIBRI is fulfilling the goal of becoming a reference tool for academia, having, as of this writing, hit the 10,000 downloads mark with users in 100 different countries. The COLIBRI platform comprises a framework—jCO-LIBRI—that provides the components required to build a CBR system. Then, several high level tools, packaged into the COLIBRI STUDIO IDE, support a novel development process for generating CBR systems based on the idea of reusing previous system designs. The COLIBRI development process is an approach for composing

---

[21] This extension is also available at http://gaia.fdi.ucm.es/research/colibri/jcolibri/contributions.

[22] http://www.cs.indiana.edu/~sbogaert/CBR/

CBR systems in a semi-automatic way. It is based on the idea of reusing *templates*, that are workflows that represent in an abstract way the behaviour of several CBR systems.

Once we have presented the platform and its development process, we have detailed the functionality provided by our tool. We have presented the basic functionality together with several complements that extend this basic functionality, namely textual CBR (see Sect. 5), recommendation (Sect. 6), knowledge-intensive CBR (Sect. 7), data-intensive CBR (Sect. 8) or distributed CBR (Sect. 9). Each extension has been illustrated with the description of a working application that uses that specific functionality.

We hope that this complete description of the COLIBRI platform may encourage readers to use our tool.

References

1. Schank, R.C., Abelson, R.P.: Scripts, Plans, Goals and Understanding: an Inquiry into Human Knowledge Structures. L. Erlbaum, Hillsdale, NJ (1977)
2. Bello-Tomás, J., González-Calero, P., Díaz-Agudo, B.: JColibri: an Object-Oriented Framework for Building CBR Systems. [74], pp. 32–46
3. Kukuric, N., Robijn, F., Griffioen, J.: The i3S document series: using case based reasoning for the solution of water stress problems. Aquastress (2008). http://i3s.aquastress.net/tools/CBR/ Aquastress I3S-Case Based Reasoning Users Guide.pdf (accessed 2012–02-06)
4. Aml-Moses project: Ambient-Intelligent Interactive Monitoring System for Energy Use Optimisation in Manufacturing SMEs (2008). http://www.ami-moses.eu/fileadmin/templates/ amimoses/files/AmI-MoSES_D7.6_EP_Platform_Public_v1.0.pdf (accessed 2012–01-06)
5. Carozzoni, J.A., Lawton, J.H., DeStefano, C., Ford, A.J., Hudack, J.W., Lachevet, K.K., Staskevich, G.R.: Distributed episodic exploratory planning (DEEP). U.S. Air Force Research Laboratory,Technical report (2008)
6. Lopes, E., Schiel, U.: Integrating context into a criminal case-based reasoning model. In: Information, Process, and Knowledge Management, eKNOW '10. Second International Conference on, pp. 37–42 (2010)
7. Serra, A., Avesani, P., Malossini, A.: Recommendation and learning. ONE Project (2008). http://files.opaals.eu/ONE/Deliverables/D4.3_SoftwareComponents_RecommenderSystem. pdf (accessed 2012–02-06)
8. Martín, A., León, C.: Expert knowledge management based on ontology in a digital library. In: Filipe, J., Cordeiro, J. (eds.) ICEIS (2), pp. 291–298. SciTePress (2010)
9. Govedarova, N., Stoyanov, S., Popchev, I.: An ontology based CBR architecture for knowledge management in BULCHINO catalogue. In: Rachev, B., Smrikarov, A. (eds.) Proceedings of the 9th International Conference on Computer Systems and Technologies and Workshop for PhD Students in Computing, p. 67. ACM, CompSysTech (2008)
10. Lotfy Abdrabou, E., Salem, A.: A breast cancer classifier based on a combination of case-based reasoning and ontology approach. In: Computer Science and Information Technology (IMCSIT), Proceedings of the 2010 International Multiconference on, pp. 3–10 (2010)
11. Díaz-Agudo, B., González-Calero, P.A., Recio-García, J.A., Sánchez, A.: Building CBR systems with jCOLIBRI. J. Sci. Comput. Prog. (Spl. Issue Exp. Softw. Toolkits) **69**, 68–75 (2007)
12. Recio-García, J.A., Bridge, D., Díaz-Agudo, B., González-Calero, P.A.: CBR for CBR: A Case-Based Template Recommender System for Building Case-Based Systems. [75], pp. 459–473

13. Recio-García, J.A., Díaz-Agudo, B., González-Calero, P.A.: Template based design in colibri studio. In: Proceedings of the Process-oriented Case-Based Reasning Workshop at ICCBR'11. (2011) 101–110
14. Recio-García, J.A., González-Calero, P.A., Díaz-Agudo, B.: Template-based design in colibri studio. Information Systems (2012) doi:10.1016/j.is.2012.11.003
15. Kolodner, J.: Case-Based Reasoning. Morgan Kaufmann, San Mateo (1993)
16. Althoff, K.D., Auriol, E., Barletta, R., Manago, M.: A Review of Industrial Case-Based Reasoning Tools. AI Intelligence, Oxford (1995)
17. Aamodt, A., Plaza, E.: Case-based reasoning: Foundational issues, methodological variations, and system approaches. AI Commun. **7**(1), 39–59 (1994)
18. Haarslev, V., Möller, R.: Description of the racer system and its applications. In: Working Notes of the 2001 International Description Logics Workshop (DL-2001), Stanford, CA, USA, Aug 1–3, 2001
19. Recio-García, J.A., Díaz-Agudo, B., González-Calero, P.A.: jCOLIBRI2 Tutorial. It/2007/02, Department of Software Engineering and Artificial Intelligence. University Complutense of Madrid (2007)
20. Shimazu, H.: ExpertClerk: A conversational case-based reasoning tool for developing sales-clerk agents in E-commerce webshops. Artif. Intell. Rev. **18**, 223–244 (2002)
21. Delany, S.J., Cunningham, P.: An analysis of case-base editing in a spam filtering, system. [74], pp. 128–141
22. Tomek, I.: An experiment with the edited nearest-neighor rule. IEEE Trans. Syst. Man Cybern. **6**(6), 448–452 (1976)
23. McKenna, E., Smyth, B.: Competence-guided case-base editing, techniques. [76], PP. 186–197
24. Brighton, H., Mellish, C.: Advances in instance selection for instance-based learning algorithms. Data Min. Knowl. Disc. **6**, 153–172 (2002)
25. Weber, R.O., Ashley, K.D., Brüninghaus, S.: Textual case-based reasoning. Knowl. Eng. Rev. **20**, 255–260 (2006)
26. Weber, R., Aha, D.W., Sandhu, N., Munoz-Avila, H.: A textual case-based reasoning framework for knowledge management applications. In: Proceedings of the 9th German Workshop on Case-Based Reasoning, pp. 244–253, Shaker Verlag (2001)
27. Lenz, M.: Defining knowledge layers for textual case-based reasoning. [77], pp. 298–309
28. Recio-García, J.A., Gómez-Martín, M.A., Díaz-Agudo, B., González-Calero, P.A.: Improving annotation in the semantic web and case authoring in textual CBR. In: Roth-Berghofer, T.R., Göker, M.H., Güvenir, H.A. (eds.) Advances in Case-Based Reasoning, 8th European Conference, ECCBR'06. Lecture Notes in Artificial Intelligence, subseries of LNCS., vol. 4106, pp. 226–240. Fethiye, Turkey, Springer (2006)
29. Recio-García, J.A., Díaz-Agudo, B., Gómez-Martín, M.A., Wiratunga, N.: Extending jCOLIBRI for textual CBR. In: Muoz-Avila, H., Ricci, F. (eds.) Proceedings of Case-Based Reasoning Research and Development, 6th International Conference on Case-Based Reasoning, ICCBR. Lecture Notes in Artificial Intelligence, subseries of LNCS. vol. 3620, pp. 421–435. Chicago, IL, US, Springer (2005)
30. Baader, F., Calvanese, D., McGuinness, D.L., Nardi, D., Patel-Schneider, P.F. (eds.): The Description Logic Handbook: Theory, Implementation, and Applications. Cambridge University Press, Cambridge (2003)
31. Díaz-Agudo, B., Recio-García, J.A., González-Calero, P.A.: Natural language queries in CBR systems. In: 19th IEEE International Conference on Tools with Artificial Intelligence (ICTAI 2007), vol. 2, pp. 468–472. Patras, Greece, IEEE Computer Society (2007)
32. Recio-García, J.A., Díaz-Agudo, B., González-Calero, P.A.: Textual CBR in jCOLIBRI: from retrieval to reuse. In: Wilson, D.C., Khemani, D. (eds.) Proceedings of the ICCBR 2007 Workshop on Textual Case-Based Reasoning, pp. 217–226, Beyond Retrieval (2007)
33. Hatcher, E., Gospodnetic, O.: Lucene in Action (In Action series). Manning Publications Co., Greenwich, CT, USA (2004)
34. Witten, I.H., Frank, E.: Data Mining: Practical Machine Learning Tools and Techniques with Java Implementations. Morgan Kaufmann, USA (2000)

35. Osinski, S., Stefanowski, J., Weiss, D.: Lingo: Search results clustering algorithm based on singular value decomposition. In: Klopotek, M.A., Wierzchon, S.T., Trojanowski, K. (eds.) Intelligent Information Systems, pp. 359–368. Springer, Advances in Soft Computing (2004)
36. Lamontagne, L., Lapalme, G.: Textual reuse for email, response. [74], pp. 242–255
37. Bridge, D., Göker, M.H., McGinty, L., Smyth, B.: Case-based recommender systems. Knowl. Eng. Rev. 20, 315–320 (2006)
38. Smyth, B.: Case-based recommendation. In: Brusilovsky, P., Kobsa, A., Nejdl, W. (eds.) The Adaptive Web. Lecture Notes in Computer Science, vol. 4321, pp. 342–376. Springer (2007)
39. Wilke, W., Lenz, M., Wess, S.: Intelligent sales support with CBR. In: Case-Based Reasoning Technology, From Foundations to Applications, pp. 91–114. Springer-Verlag, , London, UK (1998)
40. Bergmann, R.: Experience Management: Foundations, Development Methodology, and Internet-Based Applications. Springer-Verlag New York, Inc., Secaucus, NJ, USA (2002)
41. Burke, R.: Interactive critiquing forcatalog navigation in e-commerce. Knowl. Eng. Rev. 18, 245–267 (2002)
42. McSherry, D.: Diversity-conscious retrieval. [78], pp. 219–233
43. Kelleher, J., Bridge, D.: An accurate and scalable collaborative recommender. Artif. Intell. Rev. 21, 193–213 (2004)
44. Herlocker, J.L., Konstan, J.A., Borchers, A., Riedl, J.: An algorithmic framework for performing collaborative filtering. In: SIGIR '99: Proceedings of the 22nd annual international ACM SIGIR conference on Research and development in information retrieval, pp. 230–237. New York, NY, USA, ACM (1999)
45. McSherry, D.: Similarity and compromise. In: Ashley, K.D., Bridge, D.G. (eds.) Case-Based Reasoning Research and Development, 5th International Conference on Case-Based Reasoning, ICCBR. Lecture Notes in Computer Science, vol. 2689, pp. 291–305. Springer (2003)
46. Smyth, B., McClave, P.: Similarity vs. diversity. [79], pp. 347–361
47. Schulz, S.: CBR-works: A state-of-the-art shell for case-based application building. In: Melis, E. (ed.) Proceedings of the 7th German Workshop on Case-Based Reasoning, GWCBR'99, pp. 166–175. Germany, University of Würzburg, Würzburg (1999)
48. Smyth, B., McGinty, L.: The power of suggestion. In: Gottlob, G., Walsh, T. (eds.) IJCAI, pp. 127–132. Morgan Kaufmann, San Francisco (2003)
49. Recio-García, J.A., Díaz-Agudo, B., González-Calero, P.A.: Prototyping Recommender Systems in jCOLIBRI. In: RecSys '08: Proceedings of the: ACM conference on Recommender systems, pp. 243–250. New York, NY, USA, ACM (2008)
50. Quijano-Sánchez, L., Recio-García, J.A., Díaz-Agudo, B.: Happymovie: A facebook application for recommending movies to groups. 23th International Conference on Tools with, Artificial Intelligence, ICTAI'11, pp. 239–244. (2011)
51. González-Calero, P.A., Gómez-Albarrán, M., Díaz-Agudo, B.: A Substitution-based Adaptation Model. In: Challenges for Case-Based Reasoning—Proceedings of the ICCBR'99 Workshops, pp. 2–12, Univ. of Kaiserslautern (1999)
52. Díaz-Agudo, B., González-Calero, P.A.: An architecture for knowledge intensive CBR systems. [76], pp. 37–48
53. Díaz-Agudo, B., González-Calero, P.A.: Knowledge intensive CBR through ontologies. In : Lees, B. (ed.) Proceedings of the 6th UK Workshop on Case-Based Reasoning, UKCBR 2001, CMS Press, University of Greenwich (2001)
54. Díaz-Agudo, B., González-Calero, P.A.: An ontological approach to develop knowledge intensive cbr systems. In: Sharman, R., Kishore, R., Ramesh, R. (eds.) Ontologies. Volume 14 of Integrated Series in Information Systems, pp. 173–213. Springer, US (2007) doi:10.1007/978-0-387-37022-4-7
55. González-Calero, P.A., Gómez-Albarrán, M., Díaz-Agudo, B.: Applying DLs for retrieval in case-based reasoning. In: Proceedings of the 1999 Description Logics Workshop (Dl '99). Linkopings universitet, Sweden (1999)
56. Salotti, S., Ventos, V.: Study and Formalization of a CBR System using a Description Logic. [77], pp. 286–301

57. Napoli, A., Lieber, J., Courien, R.: Classification-Based Problem Solving in CBR. In: Smith, I., Faltings, B. (eds.) Proceedings of the Third European Workshop on Advances in Case-Based Reasoning (EWCBR '96). LNCS, vol. 1168, pp. 295–308. Springer-Verlag (1996)
58. Recio-García, J.A., Díaz-Agudo, B., González-Calero, P.A., Sánchez-Ruiz-Granados, A.: Ontology based CBR with jCOLIBRI. In: Ellis, R., Allen, T., Tuson, A. (eds.) Applications and Innovations in Intelligent Systems XIV. Proceedings of AI-2006, the Twenty-sixth SGAI International Conference on Innovative Techniques and Applications of Artificial Intelligence, pp. 149–162. Springer, Cambridge, United Kingdom (2006)
59. Díaz-Agudo, B., Plaza, E., Recio-García, J.A., Arcos, J.L.: Noticeably new: Case reuse in originality-driven tasks. [75], pp. 165–179
60. Aamodt, A.: Knowledge intensive case-based reasoning and sustained learning. In: Proceedings of the 9th European Conference on Artificial Intelligence—(ECAI-90), pp. 1–6. (1990)
61. Plaza, E., Arcos, J.L.: Constructive, adaptation. [78], pp. 306–320
62. Vernet, D., Golobardes, E.: An unsupervised learning approach for case-based classifier systems. Expert Update. Spec. Group Artif. Intell. **6**, 37–42 (2003)
63. Fornells, A., Golobardes, E., Vernet, D., Corral, G.: Unsupervised case memory organization: Analysing computational time and soft computing capabilities. In: ECCBR. LNAI, vol. 4106, pp. 241–255, Springer-Verlag (2006)
64. Kohonen, T.: Self-Organizing Maps, 3rd edn. Springer, Berlin (2000)
65. Witten, I., Frank, E.: Data Mining: Practical Machine Learning Tools and Techniques with Java Implementations. Morgan Kaufmann, San Francisco (2000)
66. Fornells, A., Recio-García, J.A., Díaz-Agudo, B., Golobardes, E., Fornells, E.: Integration of a methodology for cluster-based retrieval in jColibri. In: McGinty, L., Wilson, D.C. (eds.) ICCBR. Lecture Notes in Computer Science, vol. 5650, pp. 418–433. Springer (2009)
67. Plaza, E., Mcginty, L.: Distributed case-based reasoning. Knowl. Eng. Rev. **20**, 261–265 (2006)
68. McGinty, L., Smyth, B.: Collaborative case-based reasoning: Applications in personalised route, planning. [79], pp. 362–376
69. Ontañón, S., Plaza, E.: Arguments and counterexamples in case-based joint deliberation. In: Argumentation in Multi-Agent Systems, ArgMAS, Selected and Invited Papers. LNCS, vol. 4766, pp. 36–53. Springer (2006)
70. Gunderson, J.P., Gunderson, L.F., Gunderson, L.F., Gunderson, J.P.: Deliberative system. In: Robots, Reasoning, and Reification, pp. 1–17. Springer, US (2009)
71. Recio-García, J.A., Díaz-Agudo, B., González-Sanz, S., Quijano-Sánchez, L.: Distributed deliberative recommender systems. Transp. Comput. Collective Intell. **1**, 121–142 (2010)
72. Stahl, A., Roth-Berghofer, T.: Rapid prototyping of cbr applications with the open source tool mycbr. [75], pp. 615–629
73. Bogaerts, S., Leake, D.: IUCBRF: A Framework For Rapid And Modular Case-Based Reasoning System Development. Technical Report 617, Indiana University, http://www.cs.indiana.edu/~sbogaert/CBR/IUCBRF.pdf (last access: 2012–02-06) (2005)
74. Funk, P., González-Calero, P.A. (eds.) Advances in Case-Based Reasoning, 7th European Conference, ECCBR 2004, Madrid, Spain, August 30–September 2, 2004, Proceedings. Lecture Notes in Computer Science, vol. 3155, Springer (2004)
75. Althoff, K.D., Bergmann, R., Minor, M., Hanft, A. (eds.) Advances in Case-Based Reasoning, 9th European Conference, ECCBR 2008. Proceedings. Trier, Germany, Sept 1–4, 2008. Lecture Notes in Computer Science, vol. 5239, Springer (2008)
76. Blanzieri, E., Portinale, L. (eds.) Advances in Case-Based Reasoning, 5th European Workshop, EWCBR 2000, Trento, Italy, Sept 6–9, 2000, Proceedings. Lecture Notes in Computer Science, vol. 1898, Springer (2000)
77. Smyth, B., Cunningham, P. (eds.) Advances in Case-Based Reasoning, 4th European Workshop, EWCBR-98, Dublin, Ireland, September 1998, Proceedings. Lecture Notes in Computer Science, vol. 1488, Springer (1998)
78. Craw, S., Preece, A.D. (eds.) Advances in Case-Based Reasoning, 6th European Conference, ECCBR 2002 Aberdeen, Scotland, UK, September 4–7, 2002, Proceedings. Lecture Notes in Computer Science, vol. 2416, Springer (2002)

79. Aha, D.W., Watson, I. (eds.) Case-Based Reasoning Research and Development, 4th International Conference on Case-Based Reasoning, ICCBR 2001, Vancouver, BC, Canada, July 30–August 2, 2001, Proceedings. Lecture Notes in Computer Science, vol. 2080, Springer (2001)

Chapter 6
Case-Based Reasoning to Support Annotating Manuscripts in Digital Archives

Reim Doumat

Abstract The process of manually annotating ancient manuscripts may rise a problem of verifying each user annotations. Thus, it is important to use a supporting system to correct certain annotations; this may be done on the basis of the experience of other users in similar cases. In this chapter, we present how we used case-based reasoning (CBR) in a digital archive to make a recommender system, which will accelerate the annotation process and correct user errors. In fact, our system tracks important user actions and saves those actions as traces. The CBR is applied to users' traces that are considered as users' experiences. Traces are structured in reusable sections called episodes (cases); each episode represents the work done on one document unit. Episode actions comprise both problems and solutions. The recommender system compares the current episode of the annotator trace with the traces database to find similar episodes, and recommends the most appropriate actions representing the solution to the user. At that point, the user takes the responsibility of accepting or refusing the recommendations. A prototype of this application has been developed to test the performance of using the CBR system in a digital archive. The experimental results confirmed the efficiency of using case-based reasoning in this domain.

1 Introduction

In this chapter we describe how we use a case-based reasoning (CBR) approach to assist web archive users in annotating and retrieving scanned document of handwritten manuscripts, according to the work of other users. Actually, the quantity of scanned ancient manuscripts is significant and the amount of tags and annotations-added by users on this type of documents- is rapidly growing. One of the main prob-

R. Doumat (✉)
Laboratoire Hubert Curien UMR CNRS 5516, 18 Rue du Professeur Benoît Lauras, 42000
Saint-Etienne, France
e-mail: rdoumat@gmail.com

S. Montani and L. C. Jain (eds.), *Successful Case-based Reasoning Applications-2*,
Studies in Computational Intelligence 494, DOI: 10.1007/978-3-642-38736-4_6,
© Springer-Verlag Berlin Heidelberg 2014

lems with these types of documents is the difficulty to verify each user annotations or tags. Image annotation in general and manuscript annotation in particular are highly subjective tasks; the document is interpreted by humans who describe its content using textual descriptors in a given language. Image documents, beyond a certain amount, are impossible to exploit without textual descriptors. Document's textual descriptions are an extremely time consuming and subjective process. Therefore, the implementation of an annotation assistant can be very useful. These reasons encouraged us to use case based reasoning as a basic technique to the recommendation system we develop. Indeed, case based reasoning enables reusing the experience of other human users; it exploits previous encountered problems and their maintained solutions to help users determine the best choice in the current situation.

This chapter is organized as follows. Section 2 presents related works in digital archives, annotations standards, CRB and TBR, and recommendation systems. Section 3 presents an overview of our web archive as an environment to exploit CBR and assist users in annotating and retrieving scanned documents. Section 4 explains the use of CBR in our application to make a recommender system. Then, Sect. 5 gives details on our developed prototype. Section 6 shows the evaluation and the experimental results of CBR use. Finally, Sect. 7 concludes our work.

2 Background and Related Works

In this section we talk about related works concerning digital archives, annotation standards, case-based reasoning, trace-based reasoning and recommendation systems.

2.1 Digital Archives

In general, an archive is a place to store up no more used documents and ancient objects. Some archives concern the cultural heritage articles. The reserved documents are mostly unique, priceless and in restrict access. As examples of cultural heritage web archives, we mention : Digital Image Archive of Medieval Music (DIAMM) [1], Avestan Digital Archive (ADA) [2], Columbia Archives and Manuscript Collections [3], Gallica the French digital archive [4], National Audiovisual Institute website (INA) [5] and many others. These sites were created to expose and share their cultural heritage collections in form of multimedia documents. Internet users can search and visualize those documents, but they are unable to comment or annotate any image. For us, we define archive as a specialized library in cultural and historical manuscripts. Therefore, a web archive is a particular web site that exposes digital images of these rare documents. In this chapter, we propose a web (2.0) manuscript archive to facilitate accessing preciously reserved documents, as well enabling permanent and expansive manuscript search, and information retrieval. Internet/intranet users are provided with new tools to annotate these rare documents; users are traced and assisted. Annotating and extracting information from ancient manuscripts is not a recent subject [6–8]. However, manual annotation is the only technique to make handwritten manuscripts accessible and exploitable; because the use of automatic text recognition techniques gives bad results with this type of documents. Users' efforts are solicited to illuminate information about manuscript subjects and contents. Expert user annotations are expensive and difficult to obtain, for this reason we enable internet users to participate in annotating these manuscripts. The originality of our archive is in exploiting user activities to facilitate the search and the annotation processes of manuscripts' contents. Accordingly, users are traced and guided by a recommender system with CBR to correct annotations, depending on the work accomplished by other users.

2.2 Annotations Standards

In our web archive, we are interested in annotating and transcribing similarly manuscript collections, images, and image fragments. The annotation of an image fragment needs the description of the location coordinates. Moreover, users must define the access rights to the annotations that they add, each user can choose authorizing or preventing the publication of his annotations. In the future, we may need to extend annotation elements, thus the system must be dynamic enough to add new elements. Our desired annotation system should be easily transformable to other format like XML, it has to support semantic annotations and distinguish different users' annotations. Finally, users must be able to reorder manuscript pages in their personal collections.

We studied different annotation standards, which are mostly used in archives such as DC [9], METS [10], TEI [11], MARC [12], ALTO [13], in order to choose the appropriate annotation type for our archive documents. We found that METS, TEI and ALTO do not enable defining and locating a fragment in an image. DC does not offer annotation privacy at user level. MARC, METS and ALTO do not have the feature of extending cataloging elements. DC and MARC do not provide page ordering.

For all these reasons, we did not use an annotation standard, because we found that neither of them can be used alone to represent our requirements. Thus, in our archive we used a combination of the most suitable features of these standards to make our own annotation model.

2.3 CBR

CBR is an artificial intelligent approach that is used in different domains such as: legal reasoning [14], health sciences [15], cook recipes [16, 17], strategy games [18], industrial decision [19], e-learning applications [20] and web services [21, 22]. CBR is used to help users in finding a solution to their problem depending on previous registered problem-solution cases, and in making decisions. There are numerous artificial intelligent algorithms such as Bayesian networks, artificial neural networks, and rule-based systems, etc. In Bayesian networks, the task of defining the network is too complex, especially in our case where the archive contents may change with the integration of new collections and annotations. For the same reason we cannot use neural networks, because we have to connect the new node that represent any newly added item to the all relative nodes; this task is complex regarding the number of users and the added annotations.

We chose CBR because it does not attempt to generate rules or models from cases, therefore the difficulties of knowledge representation are reduced. Besides, CBR is an intelligent learning method with self-development in decision support system.

2.4 TBR

Following and registering user's interaction activities in web-based environments is called tracing, this process has always been considered as complex task. The process requires the use of methods that are capable of efficiently tracing users' activities, as well as producing tracing records that can be useful to various users. Using CBR with a base of traces- as in our work- is called Trace-Based Reasoning; this type of reasoning technique is used in dynamic systems that require adapting to changes in users' need. In such systems, traces of user interaction are considered as a knowledge source, several users' interactions might be used as experiences to solve another particular problem [23]. Actually, traces can be very resourceful for making knowledge emergence if they are registered in the system during a similar experience [24]. Different projects used TBR such as MUSETTE [25], TBS [26], and éMédiatèque [27].

In our system, our aim is to automatically retain and store the implicit knowledge generated by user interaction with the system in a trace base. The originality of our work—in comparison with the previous cited projects—is that we register the traces in a hierarchical structure in order to facilitate cases comparison and recommendation precision. We found that the representation of the traces in form of comparable and reusable episodes is important; structuring the trace in this manner permits us to extract rapidly knowledge and experience. Certainly, the trace base will grow respectively with archive use and the increasing number of new users, engendering more representative examples in the case base.

2.5 Recommendation Systems

Currently, this term expresses any system that produces individualized recommendations or has the effect to guide the user in a personalized manner to reach interesting or useful items in a huge space of choices or decisions. Our main objective is to assist archive users based on the traces of other users, to reduce time and effort needed for search and annotation processes. This is the reason why we want to use a recommender system: the intention is to support archive users to quickly reach their demands among thousands of items. Recommender systems are intelligent, they assist users in a decision making process to choose one item amongst a vast set of alternative products or services. These types of systems have been exploited in different domains especially in e-commerce to recommend users with items of interest, such as books (Amazon.com, BookLamp), photos (photoree.com), videos (youtube.com), movies (MovieLens), TV programs [28], music, art works [29], and restaurants (e.g. Entrée [30]).

In order to be operational, recommender systems must have three things: background data that are information owned by the system before the recommendation process, input data which are the information that user must communicate to the system in order to generate a recommendation, and finally an algorithm that combines background and input data to produce suggestions. In general there are two recommender techniques: basic and hybrid that mix more than one basic techniques. Using hybrid technique permits to avoid poor recommendations. Different recommendation techniques can be used relatively to system purpose; in our system we use a cascade hybrid recommender system that assembles content-based and collaborative recommender techniques, in addition to the time consideration. In a content-based system, the objects of interest are defined by their associated features (e.g. in a text using words as features). Collaborative recommender systems are the most widespread; they identify regularities between users depending on their preferences, and generate new recommendation based on the comparison between users [31]. In our archive, the use of a recommender system is important not only to recommend users to consult other archive's documents; it will also be used to correct certain annotations and avoid repetitive errors.

3 Web Archive of Scanned Handwritten Manuscripts: An Application to Search Information and Annotate Online Manuscripts

In this section, we describe the general structure of the archive with its two main parts: the first one concerns the documents and their organization; the second one explains the tracing and recommending systems.

3.1 Documents' Structure in the Archive

We will detail the archive structure because cases in CBR are built upon archive elements and users' activities that generate traces. Users' traces will be treated in

Sect. 3.2, where we explain the way we generate episodes (cases) from traces depending on the manipulated archive document unit. Currently, we define the difference between our archive document units (archive contents), and show how they are organized to facilitate their classification and annotations.

3.1.1 Collections

A collection comprises scanned pages of a handwritten book or some related documents; it contains more than one image (page). Images' order reflects the original order in the paper collection. *Original collections* are added by the archive administrator while *Personal collections* are created by end users. Personal collections enable users to gather several images from diverse collections into new ones with different image order. Additionally, they enable users to express their point of view about manuscript subjects depending on their experience in the domain. Users can manage their personal collections (create, modify, delete, and add images), and define the access rules.

3.1.2 Images

Each image represents a scanned page of the manuscript document. Images belong to one original collection in order to be introduced into the archive for the first time. Then, images can be gathered or separated in new personal collections.

3.1.3 Image Fragments

Image fragments are concave closed shapes drawn on an image. Each fragment is defined by a set of points. Point coordinates (x, y) are expressed in relation with image dimensions where fragments were created; this manner allows fragments to appear always at the same position in the image even when image is resized. In general, image fragments will be visible to all users unless the annotator (the user) defines his annotation as private.

We consider collections, images, and image fragments as document units. Thus, a *document unit* is an abstract element that represents any type of document or document part. A document unit may contain another document, for example, a collection contains images, and an image may hold different fragments. The main objective of introducing the document unit is to simplify annotation anchoring, because annotations can be added only on document units. In Fig. 1 we present an example about document units' representation. The collection "Bolyai notes 01" is a document unit, as well as the image (BJ-563-4.jpg) in this collection, and the fragment (Frg_26) that is a part of the referenced image.

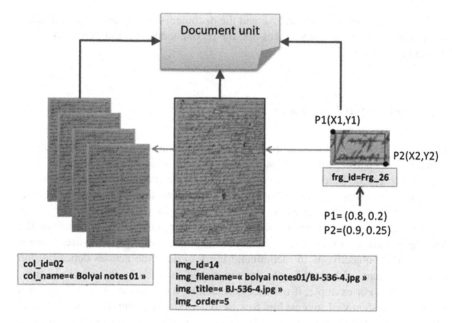

Fig. 1 An example of document units

3.1.4 Annotations

In our archive model, annotations represent words added by any user on any document type (collection, image or fragment). They are rather uncontrolled vocabularies that express a personal point of view, a script transcription, or some relative information about the manuscript subject. In fact, we aimed to provide annotation tools that enable users to work in collaboration, to annotate a defined area of an image, and to define their annotation confidentiality. Additionally, we intend to maintain annotation evolution for new users' requirements. Therefore, our archive annotations should be easily converted into XML based syntax in order to simply change them into other types of metadata.

Each annotation in our archive is composed of *annotation type* and *annotation value*. As for the Annotation type it will be defined by the archive administrator (comments, questions, answers, keywords…). Presently, in our prototype we use two annotation types, the first one is *Keywords* signifying any free word or text added by the user, and the second one is *Transcriptions* that must designate the same letters in the defined fragment. Annotation values are the vocabularies added by users, these later may also define the language of their annotation values. This might be helpful to filter the search results depending on a given language in case of archive collections are in different languages.

We presented till now the archive and the documents' structure. In the following paragraph we describe users' traces and the recommender system that is based on the users' activities.

3.2 Tracing and Recommending Systems

Due to the fact that our web archive represents a collaborative work environment, different users participate to annotate a document unit, which could be a collection or an image fragment. Some users may make errors while annotating these documents; possibly other users will try to correct the annotations that they consider inaccurate. Thus, the use of a tracing system to capture user's actions and experiences is essential; first to recognize the work of every user, and second to distribute the knowledge between users. Above and beyond capturing and registering user actions, the important objective is to recommend users to do some actions depending on their last actions, and according to the experience captured by the system in similar situations. For example, if the user is searching a document, besides his search results, the system will suggest him to view other relative documents seen by other users who already did similar search. If the user is modifying or correcting registered annotations, the assistant will capture these actions and register them for further use to prevent repetitive errors in annotations. All the suggestions are generated by the recommender system. Therefore, tracking user actions could increase the annotation quality and the recommender system performance. The recommender system in our web archive is built on users' traces and artificial intelligent technique of type case-based reasoning; where traces of users' activities will be cut into small parts (episodes) to form a container of cases. While users interact with the system, browse documents, compose new collections, add annotations they leave many traces in the trace database as shown in Fig. 2. Thus, the recommender system will use CBR and take into consideration the user profile to propose recommendations and answer the question: "What to do now, according to the user last action and the traces of other users in similar situations?". A case-based reasoning system (CBR) is an assistant for situations that are difficult to formalize such as using a web archive. CBR keeps a number of different knowledge containers (case base) of problem-solving process that will be used to solve a new problem. Thus to use CBR in our archive, we consider users' activities about browsing and annotating images as a case base; the last action of the current user is the problem to be solved, and the system recommendations as solutions. In our application, users do not offer a feedback to the intelligent system, however the system can estimate from user traces if he took the recommendations in consideration or not.

In the following sections we present the three main parts used by the case-based reasoning system in our web archive.

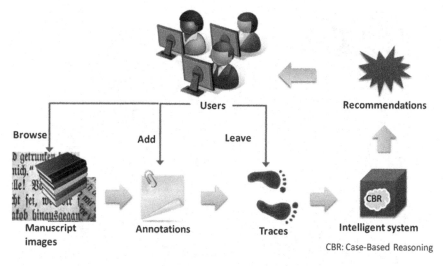

Fig. 2 A global view of our web archive components and their relationships

4 Case Based Reasoning

We use CBR in the web archive to facilitate experience reuse in repetitive tasks like searching relative documents and verifying added annotations. The basic idea of using CBR is to exploit precedent encountered and resolved cases to solve a new problem, first by identifying the resemblance between the novel case and the ancient ones, then by choosing the most appropriate solution from the earliest resolved cases. The principle is that similar problems have similar solutions, thus collected solved cases help in making new decision. Additionally, CBR eases learning from experiences without the need of detailed modeled domain.

An example of the desired suggestion function of our system can be illustrated in Fig. 3, a user annotated a fragment with the word "nen" (1), and then another user corrected the content of the annotation into "nem" (2). Now, when a third user tries to add the word "nen" once again into a fragment (3), the system will suggest the user to change his annotation into "nem" since it was corrected somewhere. The problem when working with handwritten documents is that there is not a technique that permits to verify the annotation added by users. Thus, identifying an error will remain a subjective topic.

In this section, we demonstrate how to build cases from the interaction between the users and the system. Then we explain how we retrieve similar cases to adapt a solution, this later will be represented in form of suggested action to the current user.

Fig. 3 System suggestion based on registered corrections

4.1 Building Cases Base from Users' Traces

Case generation is an important part, according to case structure we can build fast
retrieval strategy or not. Following and registering user's interaction activities in web-
based environments is called tracing. In our archive, we use hierarchical structure
to register cases that are trace's chunks. Firstly, we register the user actions while
handling the archive document units. The registration during a session is considered
the whole trace. In order to build the case base, we divide the trace actions into
small groups that are called *episodes*; each group concerns the work done on one
document unit (depending on the *id* of the document unit). The division process is
done simultaneously with the trace registration. Thus these episodes are the cases
in our system. We do not separate the actions in each episode into problem part and
solution part, because when we compare the current user episode with the registered
ones in the case base, we consider that the last user action is the *problem part* in
the similar registered episode, and the *solution part* that we look for will be all the
followed actions in this episode. The actions' order is very important because we do
not want to recommend the user with some actions that he just finished doing.

As our intention is to observe user's activities only on the server side (searching a keyword, annotating a document, etc.); user's interaction on the client side (using scroll bars, forward, backward, mouse click…) is not completely traced. Before ongoing in explaining the traces and their use in cases, we give some definitions.

4.1.1 Trace (T)

In our model, a *trace* is a record of the user interaction with the manuscript archive throughout a working session. A working session begins with a login to the application and ends with the log out. Thus, a trace represents a sequence of user actions (actions will be detailed in the next Sect. 4.1.2) as mentioned in the equation below.

$$T = \{a_k, k = 1..n\}$$

where: a_k is an action
K represents the order of the action (a) in the trace.

An example of a user trace is illustrated in Fig. 4; the trace always begins with the action a_1 "Login". User trace reveals the accomplished work and the documents of user interest.

4.1.2 Trace Components

Action is a trace basic unit; user actions may modify the archive documents when user creates a fragment or adds an annotation. Some actions do not change the archive contents such as selection. Trace actions can be repeated as the user performs the same action on different document units. Before proceeding and explaining the trace structure we will identify actions and their components.

Action (a): in our concept, an action is an event made by the user that affects a document unit during a session. Each action is composed of a type (*AType*) and a set of parameters (*Pi*)

$$a = (AType, \{P_i\}_{(i=1,m)})$$

where: m is the number of the action parameters; it is relative to the action type.

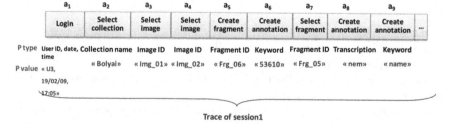

Trace of session1

Fig. 4 An example of a user trace

Action type (AType): action type defines action category and directs the system how user exploits archive documents. In the archive, users perform different types of actions, we do not trace all their actions; because we are only interested in actions about search, browse and annotation process (create fragments and annotations). Thus, we trace the following action types: Login, Logout, Select, Add, Create, Delete, and Edit (Modify). These actions are done with all document units' types; therefore we define Select Collection, Select Image, Select Fragment, Select Annotation, Add Collection, Add Image, Add Annotation, etc.

Action Parameter (p): action parameters are the items that are affected by an action and they differ depending on action type. For example, the first action "login" (in Fig. 4) has three parameters which are: (User_id, date, time), the second action "Select collection" has two parameters (Collection ID, Collection name), the action "Create fragment" has six parameters (Fragment ID, Image ID, X, Y, width, and length).

Each parameter has a *type* (PType) representing the name of the parameter, and a *value* (Pvalue) holding the data in the parameter.

$$p_i = (PType, Pvalue)$$

For example, in Fig. 4 the action "login" has (user_id, date and time) as parameter types and (U3, 19 Feb 2009, 17:05) as values of these parameters.

Actually, traces are hard to be exploited (compared and reused) when they are long sequences of actions. Therefore, decomposing traces into smaller reusable units that we call *episodes* (see Sect. 4.1.3) will be very useful to the recommendation system. In fact, the tracing method affects the recommendation type and in our archive the recommender system generates recommended actions instantly. Thus, it must rapidly find similar actions and calculate the most appropriate action for the user. Therefore, structuring user traces in form of episodes will facilitate their exploitation and reuse.

4.1.3 Trace Episodes

We use the word *episode* to describe a chronologically consecutive cluster of actions that concern the work done on one document unit. An episode *ep* is defined in the following equation.

$$ep = \{a_i, i = 1..n\}$$

where:

    a_i Episode actions ordered chronologically

    n is the number of actions in the episode

The episode length is not limited, it depends on the number of actions that the user done on the document unit.

An example of trace episodes is illustrated in Fig. 5 The episode ep_1 concerns the actions on the collection (Bolyai), the episode ep_1.2 contains the actions that affect the image (Img-02). As we can see in this figure, the episode ep_1.2 composes a part of ep_1, this lead us to define the sub-episodes.

4.1.4 Sub-Episode

When an episode ep_2 forms a subset of actions from another episode ep_1, we say that ep_2 is a sub-episode of ep_1. Likewise, ep_1 is considered as a parent episode of ep_2. This happens when the document unit (du_2) of the sub-episode ep_2 is contained in the document unit (du_1) concerned by the parent episode ep_1. Thus, we define:

$$du_2 \subset du_1 \Leftrightarrow ep_2 \subset ep_1$$

Furthermore, we define two types of episodes in our archive:
Simple episode, is an episode that does not have any sub episodes, for example the episodes (ep_1.1, ep_1.2.1, and ep_1.2.2) in Fig. 5.

Complex episode contains at least one sub-episode, for example (ep_1) in Fig. 5 we show a complex episode; it represents the work done on a collection and its images.

4.1.5 Episode Hierarchical Structure

Episodes represent actions carried out by a user on the same document unit, thus good organization of these episodes in the system storage space easily leads to retrieve common features between users' work.

As we illustrate in Fig. 5, when the user changes the document unit, the tracing system will begin a new episode. Considering that a document unit may represent one of the following types of documents (collection, image or fragment), registering the episode in the database requires identifying if the episode is simple or complex.

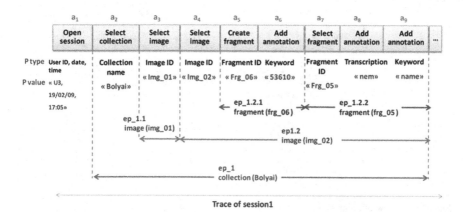

Fig. 5 The episodes and their hierarchical structure in a trace

Complex episodes have a hierarchical structure (levels) according to the document unit type. For example in Fig. 5, the first level (collection level) concerns the actions done on collections (ep_1), second level (image level) represents actions done on images (ep_1.2), and the third level (fragment level) corresponds the actions of image fragments (ep_1.2.1, ep_1.2.2).

The hierarchical structure allows episodes to be assembled into meaningful compositions, and provide the basic structure needed for quick and relevant retrieval of these episodes. The hierarchical structure will be useful for the assistance to extract episodes of the same level to the current user actions, in order to generate useful recommendations.

After registering and structuring traces in the database, the next step in the system is to compare current user trace with the registered episodes of the same level to find similar ones. The system will continuously compare the last unfinished episode of the current user with the episodes in the trace database. The challenge is to recommend the user in the real time depending on his last action. For this reason we do not wait the user to finish his session. The problem part for the system will be the last action a_k of the current user (Fig. 6), and the solution that the system is seeking will be the most suitable next registered action among all the next actions that it will find in the case base (of course for only the similar episodes).

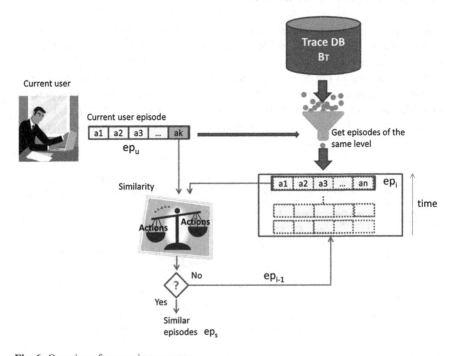

Fig. 6 Overview of comparison process

4.2 Similarity Matching for Case Retrieval

According to the idea of Case Based Reasoning, retrieve the most similar case with respect to the current user's work might help him in solving his problem which is in our application the next action. This raises the question of detecting similar cases *"Which characteristics should be taken into account to determine similarity between cases?"*. The answer of this question cannot be general because each algorithm has to be adapted to the system requirements. In our system, as a result of the proposed hierarchical structure of user traces and their division into episodes depending on the document units, the system compares registered episodes with the current user episode rather than comparing whole traces. Furthermore, the trace hierarchical structure permits to only compare episodes of the same type (same level), thus to reduce the mining time and to merely propose relative actions to the user. We proposed the following algorithm (Algorithm1) which extracts similar cases (episodes) and calculates the similarity degree depending on episodes' actions and their parameters. Figure 6 displays a global view of our algorithm. Once we identified the most alike situation we can use another algorithm to recommend the user to do actions based on the results of Algorithm 1. For example, if we observe the trace in Fig. 5, the last unfinished episode of the highest level will be the episode (ep_1.2.2) of fragment level, the system will extract all episodes of fragment level to measure their similarity with the current user episode.

Following, we describe the functions used in the algorithm.

Algorithm 1: Extracting similar episodes
---

1. **Inputs: user's last unfinished highest level episode**(ep_u)**, trace database**(B_T)
2. **Method:**
3. **From**(B_T)**, get all episodes** (ep$_i$)**of the same levelas**(ep_u)**, in inverse chronological order.
4. For all returned(ep$_i$),**start the comparison between episodes**(ep_u) and (ep$_i$) , from the most recent till the oldest ones.
5. **Compare** (ep_u) with (ep$_i$) to calculate their similarity degree depending on the similarity of their actions*(details cf.§ 4.2.2)*
6. For all actions (a_u) in(ep_u)
7.   For all actions (a_i) in (ep$_i$)
8.   Compare the similarity between these actions and action (a_u) *(details cf.§ 4.2.2)*
9.   For i=1 to $|ep_i|$ (the number of the actions in the episode ep$_i$)
10.   **Get the max of similarity:** $sim(a_u, a_i) = $ **max.sim**(a_u, a_i)
11.   End For
12.   End For
13. End For

4.2.1 Finding Episodes of the Same Level

The first step in finding similar episodes is to filter the traces in the database (B_T) to get all episodes (ep$_i$) that have the same level as the last unfinished episode of the current user (ep_u). This is achieved by calling the function $f_{simLevel}(ep_u, B_T)$,

$$f_{simLevel}(ep_u, B_T) \rightarrow \{ep_i, i = 0, k\} \tag{1}$$

Where k is the number of the returned episodes

ep_i is an episode of the same level as ep_u

Now for all returned ep_i the system calls the function of comparing ep_u with these episodes starting from the most recent episodes.

In what follows, we define a simple comparison method for episodes. We agree that more complex algorithms could be used, such as an adaptation of the BLAST algorithm (Basic Local Alignment Search Tool.[1]) BLAST is used to rapidly compare a given gene sequence X to the sequences in a database to find all similar ones.

4.2.2 Comparing Two Episodes of the Same Level

In order to measure the similarity between two episodes (Line 5 of the Algorithm1), the system calls the function Sim_{ep} between two episodes e_1 and e_2, where e_1 represents ep_u and e_2 represents an ep_i

$$sim_{ep}(e_1, e_1) = \sum_{i=1}^{|e_1|} f_{max}^{|e_2|} (Sim_{Action}(a1_i, a2_i)) \tag{2}$$

where $a1 \in e_1$ and $a2 \in e_2$ with i: $1..|e_1|$ and j: $1..|e_2|$ and $|e_i|$ is the number of the actions in the episode e_i. (This number is not constant and differs from episode to another) and the function registering the max of $sim(a_u, a_i)$ (line 10 of the Algorithm1).

The order of the actions in the episodes is important, because once the function sim_{ep} gets the max of similarity between two actions (when comparing two episodes); it needs to mark the next action in e2, which will be used later in the recommendations.

The function $Sim_{ep}(e_1, e_2)$ will call in its turn Sim_{Action} to calculate the similarity between two actions a1, a2. We need to mention that if Sim_{Action} is less than a specified threshold, it will be considered as 0. This threshold is used to assure the quality of the recommendations. It was defined by our experimentations to 0.7 (see experimental results Sect. 6.3). The range of sim_{ep} is between zero and the number of the actions in the episode ep_u.

---

[1] http://en.wikipedia.org/wiki/BLAST#BLAST

Table 1 Table of similarity between action types

Action type	Login, logout	Select	Create	Delete	Edit (modify)
Login, logout	1	0	0	0	0
Select	0	1	1	0	0
Create	0	1	1	0	0
Delete	0	0	0	1	0
Edit (modify)	0	0	0	0	1

4.2.3 Comparing Two Actions to Measure Their Degree of Similarity

Calculating the similarity between two actions requires the results of two other similarity functions, the first is to determine the resemblance of the two actions' types (*AType*), and the second is to estimate the similarity between the actions' parameters (P).

Before calculating Sim_{Action}, we get the number of their parameters, if the two actions have the same number of parameters we continue to calculate the similarity Sim_{Type} and Sim_{par}. Otherwise, $Sim_{par} = 0$, and the similarity between actions equals zero.

The similarity between two actions a1, a2 is calculated by the formula:

$$Sim_{Action}(a_1, a_2) = Sim_{Type}(AType_1, AType_2) * \frac{\Sigma_{i=1}^{m}(P1_i, P2_i)}{2m} \qquad (3)$$

Where: m is the number of parameters of the action a1 and a2
Results from Sim_{Type} is in the range [0,1].

We divide Sim_{Par} by 2m because the similarity between two parameters (Eq. 3) is a result from the addition of two similarities and Sim_{ptype} and Sim_{pvalue}, thus it will be in the range [0,2]. In Eq. (3), the division by 2 permits to get a similarity Sim_{Action} between 0 and 1.

Comparing Two Action Types

The similarity between the two action types is given by Table 1. We consider that actions' types *Select* and *Create* are similar. Currently, it does not make sense in our work to compare actions of different types.

If $Sim_{Type} = 0$ then Sim_{Action} will be zero also, we do not continue to calculate the similarity between their parameters. Else we compare the parameters of the two actions. The number of the parameters differs between two actions even if both have the same type. For example, the action *Create* for a fragment has six parameters and *Create* for an annotation has three parameters. Thus, if the number of the parameters of the two compared actions is not the same, we do not continue the comparison and consider that the similarity between the actions equals zero even if the action type is the same.

Comparing Two Action Parameters

For two actions of the same type, the parameters are stored in the identical order. Consequently, we can couple the parameters from the two actions in order to compare their types and values:

$$Sim_{par}(P_1, P_2) = Sim_{ptype}(PType_1, PType_2)$$
$$+ Sim_{pvalue}(PValue_1, PValue_2) \qquad (4)$$

Calculating the similarity between two parameters is done by the addition of the results from two functions Sim_{ptype} that is given in the next formula and Sim_{pvalue} that is given in Eq. (6). In Eq. (4), we used the addition (not the multiplication) to get a positive degree of similarity if the two parameters have the similar values and dissimilar type instead of zero. As in the case of two annotations that have the same value "nen", but they have different types (keywords and transcription).

$$Sim_{ptype}(PType_1, PType_2) = \begin{cases} 0 & if \ PType_1 \neq PType_2 \\ 1 & if \ PType_1 = PType_2 \end{cases} \qquad (5)$$

If Sim_{ptype} is zero, we do not compare their *pvalue*, otherwise we calculate the similarity between the *pvalue*:
Sim_{pvalue} is calculated according to the distance between the two parameters,

$$Sim_{pvalue}(PValue_1, PValue_2) = 1 - distance \qquad (6)$$

The distance is measured according to parameter type, for parameters that hold annotations (strings of characters); we measure the distance using Levenshtein algorithm (also called edit distance[2]). The distance is defined as the minimum number of edits needed to transform one string into the other. Here a comparison function using semantic resources can be also used. Parameters of number type, such as the coordinates of fragment points, the distance is given by the formula:

$$distance = \frac{|PValue_1 - PValue_2|}{PValue_1 + PValue_2} \qquad (7)$$

For parameters of id type, the distance is calculated with the formula:

$$distance = \begin{cases} 0 & if \ PType_1 \neq PType_2 \\ 1 & if \ PType_1 = PType_2 \end{cases} \qquad (8)$$

---

[2] http://en.wikipedia.org/wiki/Levenshtein_distance

For example, two actions concerning fragments could be similar if they have similar annotations (transcriptions or keywords), or if they concern very close fragment coordinates in the same page. If the actions handle pages then they could be similar if they have similar annotations.

An example about this algorithm is illustrated in Fig. 7.

In the example, we realize that the filter gives out the episodes (ep_1, ep_2, ep_3; which are of the same type "fragment episode" as the current user episode ep_U. After comparing the user episode with the three resulted from the filter, the algorithm will retain only two episodes (ep_1 and ep_3) that are similar to the user one with a similarity degree $>= 0.7$.

To explain how we got the results in Fig. 7, we will use the algorithm equations with the example illustrated in this figure to expose the result of the comparison between ep_u and ep_3 in Table 2, we show the end result of the comparison between all of two episodes' actions:

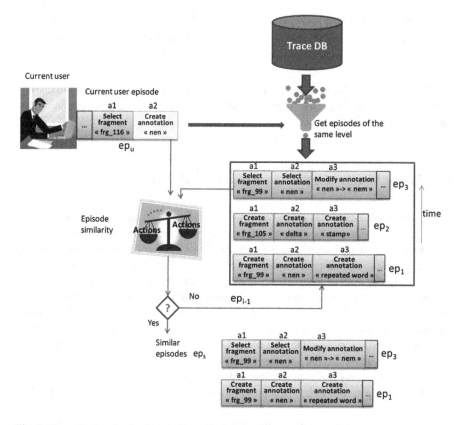

Fig. 7 Example about extracting similar episodes algorithm

Table 2 Calculating the similarity between ep_u **actions and** ep_s **of the example in** Fig. 7

Action	Action	Sim_{Type}	Sim_{ptype}	Sim_{pvalue}	$sim(a_u, a_i)$
a1	a1	(Select, Select) = 1	(Frg_id,Frg_id) = 1	(116, 99) = 0	0.5 → 0
a1	a2	(Select, Select) = 1	(Frg_id,annotation) = 0	–	0
a1	a3	(Select, Modify) = 0	–	–	0
a2	a1	(Create, select) = 1	(annotation, Frg_id) = 0	–	0
a2	a2	(Create, select) = 1	(annotation,annotation = 1)	(nen,nen) = 1	1
a2	a3	(Create, modify) = 0	–	–	0

Thus, upon the results of this table, the episodes (ep_u) and (ep_3) have two similar actions those are the second actions. Likewise, the algorithm will find other similar episodes to (ep_u).

Therefore:

$$is\ a1\ similar\ to\ an\ action\ in\ ep3 = max(sim(a1, a1), sim(a1, a2), sim(a1, a3)) = 0$$
$$is\ a2\ similar\ to\ an\ action\ in\ ep3 = max(sim(a2, a1), sim(a2, a2), sim(a2, a3)) = 1$$
$$sim_{ep}(epu, ep3) = 0 + 1 = 1$$

The resulted similar episodes will be the input of the recommender system (assistant), which decides the most suitable next action to the current user depending on the degree of similarity between the episodes.

4.3 Reuse Similar Cases in a Recommender System

Reuse the similar extracted cases from Algorithm1 to solve current user problem is done by the recommender system. Algorithm2 represents the work of our recommender system, which mainly takes as input the user's current episode and the results of the episodes' similarity from the first algorithm. The retrieved similar episodes are used by the system to decide the best next action for the user. The suggested action is issued after the calculation of the most similar situation to the current user context to give accurate recommendation. The output of Algorithm2 could be zero or several actions (ordered in priority). The suggestion list will be empty if no match is found for two reasons, either the trace database is empty or the similar episodes in the trace database are not sufficient.

As we can see in Fig. 8, the recommendation algorithm has three inputs: the user profile (working groups), the current user episode (ep_u), and a set of similar episodes (ep_s).

We present the recommendation algorithm 2 and the details will be explained in the following paragraphs.

Algorithm 2: Recommendations

---

1. **Inputs:** current user profile(pf_u), current user episode (ep_u), list of similar episodes (Ep_s) from Algorithm1.
2. **Method:**
3. $(a_u) \leftarrow$ *last action of* (ep_u)
4. Extract the following actions (A_n) for all *similar episodes* $(ep_s \in Ep_s)$:
5. From each (ep_s), get the action (a_n) that follows the similar action to (a_u) in order to get the pair (ep_s, a_n)
6. Create a list of episodes with their associated next action : $List(ep_s, a_n)$
7. Reverse the list to get next actions with their associated episodes $List(a_n, List(ep_s))$
8.   For all (a_n) in $List(a_n, List(ep_s))$
9.   For all (ep_s) in $List(Ep_s)$
10.   Measure profiles similarity sim_{pf} between current user and the user of the registered episode.
11.   Get episode similarity sim_{ep_s}
12.   Transform the episode time into percentage score (t)
13.   Calculate the recommendation rate $r(a_n)$
14.   End For
15.   Calculate the (a_n) recommendation rate $R(a_n) = \sum r(a_n)$
16.   End For
17. Decide the actions (a'_r) to be recommend from (a_n) that have greatest recommendation rate and make the list of recommendation actions R

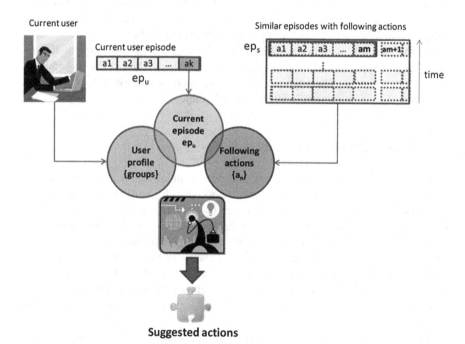

Fig. 8 A global view of the recommendation system

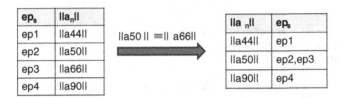

Fig. 9 Converting the list of similar episodes and their following actions

4.3.1 Extracting the Following Actions (a_n)

This function acquires as parameters a set of similar episodes (Ep_s) to the user's one (ep_u), it locates the similar actions to the last user action (a_u), and returns a set of actions (A_n) that follow the similar actions in (Ep_s) (Fig. 10). The recommender system handles these actions (A_n) to produce qualified suggestions to the user work. The following actions (A_n) could be found in the similar episodes (Ep_s) or in the episodes that follow them.

The system creates a list of following actions (a_n) with their associated episodes (ep_s) by reversing the list of (ep_s, a_n). For example, if we have a list of similar episodes with their next actions as in the left table in Fig. 9, after the reversion the results will be as in the table to the right. Where $\|a\|$ is a norm of the action calculated using the action type, parameters' types, and parameters' values.

By this function, we consider that two following actions are identical if they have the same type, the same parameter types and values. Thus in Fig. 9, $\|a50\| = \|a66\|$ have the same action type, parameter types and parameter values, and they are represented as one action $\|a50\|$ in the right table.

4.3.2 Measuring Profiles Similarity

In our archive, we have administrators and users. Administrators are responsible of adding new collections to the archive, create accounts for new users, grant permissions to users, create new groups and assign users to groups. Users have only two permissions: *Read,* which means access the exposed documents units, and *Write* if they are allowed to add and modify document unit's annotations. When a new user account is created, the user is assigned to some groups (Historians, Linguists, Librarians, Hungarians, French, Students…) and his profile will be composed of these groups. The interest of taking into consideration the user profile is to recommend the user with document units that may concern him depending on the interest of his groups.

The Sim_{pf} function calculates the similarity between two user's profiles, a user profile is defined by the groups that the user belongs to. The Sim_{pf} will be used later in the recommendation rate equation, to give more importance to the episodes of users who are similar to the current user profile.

$$Sim_{pf} = \left(\frac{A \cap B}{A \cup B} \right) \tag{9}$$

Where A and B represents sets of user groups. For example, the current user "David" belongs to the group (Hungarian manuscripts), and the compared trace in the database is left by the user "Abel" who belongs to the groups (Hungarian manuscripts, Hungarian linguists). Thus between "David" and "Abel" is:

$$Sim_{pf} = \left(\frac{Hungarian\ manuscripts}{Hungarian\ manuscripts,\ Hungarian linguists} \right) = 0.5$$

4.3.3 Transforming Episode Time into Score

The date and time when the episode was created is important in the recommendation process. Recent episodes are preferred more than old ones for the reason that document units or their parameters might be modified with the time. Thus recommending users with up to date parameters is more useful than recommending expired parameters that are no more used in the archive.

To integrate the time aspect in the recommendation, we propose to transform it into a score (t). Next we give the formula to do this transformation.

$$score(t) = k/(k + 10 * (Current\ date - registered\ episode\ date)) \tag{10}$$

Where: k represents the number of acceptable days before considering (t) score as negligible ($<10\%$), it is defined by the system administrator depending on the frequency of application use. It is merely linked to the number of archive users.

We remind the reader that our system recommendations do not affect the existence of the document units (collections, images, image fragments). For example, if an episode about rare accessed collection is registered many years ago, and the system does not recommend it anymore because of the score(t), the user may find this collection by a simple search and his registered trace will bring back the collection to the recommendations.

4.3.4 Calculating the Recommendation Rate

This is the last function in the algorithm that will calculate the recommendation percentage of every candidate action (a_n) as specified in the formula (11) according to weighted hybrid recommender systems:

$$r(a_n) = (C_1 * sim_{pf} + C_2 * sim_{(ep_s)} + score(t)) \tag{11}$$

Where C1 and C2 are weighted coefficients used to define the importance of $sim_p f$ and $sim(ep_s)$. In our system, C2 > C1 > 1 because we consider that $sim(ep_s)$ is more

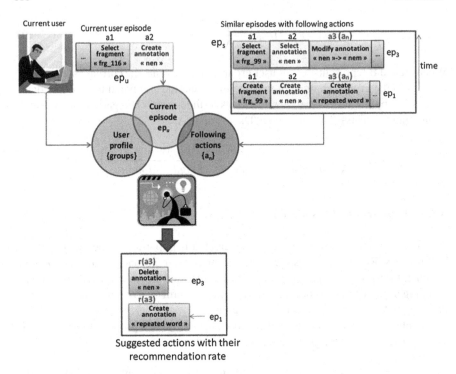

Suggested actions with their
recommendation rate

Fig. 10 A simple example about recommendation algorithm

important than $sim_p f$ which is more important than score(t). During the experiments we found that the use of $(C2 = 7, C1 = 3)$ and $(C2 = 8, C1 = 2)$ advises suggestions that are not adapted to the current user profile; for example to consult some collections that are not of his language or interest. For that reason we used $(C2 = 6, C1 = 4)$, which gives more acceptable results for the user profile.

After computing recommendation rate for each action separately, this rate will be assembled for the same actions to give a higher recommendation rate for that action.

$$R(a_n) = \sum r(a_n) \tag{12}$$

Finally the recommended actions will be sorted from the higher to lower rate, in order to make more relative suggestions to current user actions. Next, Fig. 10 gives an example about the recommendation algorithm; it takes as input the results of the example in Fig. 7, which are (ep1, ep3), the user profile (his groups), as well as the current user episode (epU) that contains in our example two actions a_1(Select fragment 116) and a_2(Create annotation "nen").

Then, the recommender system will mark—in each similar episode—the most similar action to the last action of current user (a_U), which is (Create annotation "nen" at the user side) in Fig. 10. In this example, the similar actions are a_2 (Cre-

Table 3 Calculating the recommendation score for suggested actions of Fig. 10

Similar episode eps	Similar action to a_u	Following action a_n in eps	sim_{pf}	sim_{eps}	Score(t)	$R(a_n)$
ep3	a2	a3	0.5	1	0.3	8.3
ep1	a2	a3	0.5	1	0.1	8.1

ate annotation "nen") in ep_1 and a_2 (Select annotation "nen") in ep_3. Afterward the algorithm takes the next action (a_n) for each marked action, that means it will take the actions a_3 (create annotation "repeated word) from ep_1, and a_3 (modify annotation "nem") from ep_3. Then it will calculate the most suitable next actions to be recommend to the current user.

Following, we describe in details the functions used in the algorithm to calculate the recommendations.

To explain Fig. 10, we will use the described equations with the example illustrated in this figure. The recommendation algorithm extracts from the similar episodes - resulted from the similarity algorithm (ep3, ep1)—the actions similar to the last action in ep_u. In Table 3, we show the scores of the recommended actions of the example in Fig. 10.

The recommendation rate is computed with $C1 = 4$ and $C2 = 6$.

The recommender system generates relative suggestions, and delegates the responsibility to the user for making the final decision, which is to take the suggestion into consideration or not. In both cases, the system will retain the decision of the user in form of trace that will create a new solved case, and later this trace will be taken into consideration by the assistance in upcoming recommendations. Furthermore, we believe that our system has an auto-correction function, because the suggestion that is not used by any user (a wrong one or a non relative suggestion) will not be proposed frequently in the future, this is relative to the variable *score(t)* in the Eq. (11) that gives the importance to the recent episodes to be recommended.

5 Web Archive Prototype with CBR in Support to Annotate Ancient Manuscripts

We have developed a prototype for digital manuscript archive called ARMARIUS. The aim of this system is to disseminate images of handwritten documents in a Web 2.0 application with appropriate tools to annotate any document type, as well as to assist users in the annotation and the research process of these kinds of manuscripts. The archive provides some assistance tools to show every user a list of his actions and a list of proposed actions. Archive users can inspect documents/images, construct personal collections, add annotations onto images and collections, and create fragments on images to add more specific annotations or transcriptions. Furthermore, the archive displays the trace of current user inside a widget at the top of the working space; it also displays an assistant window presenting several suggestions according to current user actions, as in Fig. 13.

5.1 System Architecture

ARMARIUS is developed in Java using the MVC2 design pattern. The system architecture and the development technologies are illustrated in Fig. 11. The main part of our application is hosted on a web server; we used GlassFish as a web server with JSF framework to develop the application. The JSF framework is mapped with the database via persistence layer that uses Hibernate. The database holds manuscript images, all type of annotations, as well as traces of users' interaction with the system. We used MySQL to manage the database with archive elements, the database can be hosted by the same application server or another server connected to the web server. When a user interacts with the application through a web browser as shown in Fig. 11 (at the right), the application authenticate the user and verifies his rights. Then it executes the request action and, if necessary, reads and writes data in the database. At the client side, we use JavaScript and Ajax to interact with the web server.

5.2 Annotation Tools

In ARMARIUS users can annotate any type of document units (collections, collection page, and fragments). The annotation of collections and pages is similar, users choose the document to be annotated and click on the annotation tool, an annotation window appears permitting the user to edit his text and the annotation type. Annotating a fragment is slightly different from annotating collections and pages, because users have to draw the fragment (a closed shape) on the image before adding their text. Users can drag the drawn fragment to another position in the image, adjust the shape dimensions, choose the annotation type (keyword, transcription), add words, or delete the fragment. Furthermore, the annotation tool maintains and shows the user name who added the annotations. An example of the annotation process is shown in Fig. 12.

5.3 Tracing System to Generate and Register Cases

The trace repository was implemented as a centralized database server with MySQL. The choice of using a relational database has several advantages: traces are structured in a rich format, they can be easily restructured and transformed into another format (e.g. XML/RDF or text), moreover it enables trace manipulation such as insertion, cut, etc.

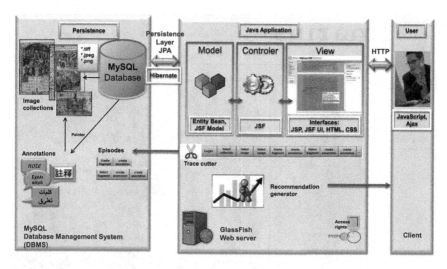

Fig. 11 ARMARIUS architecture with the used technologies

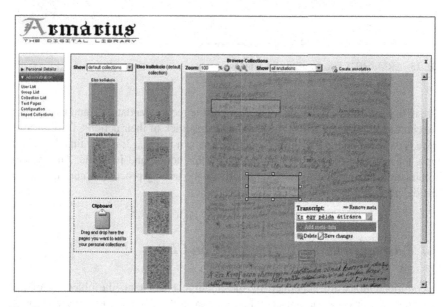

Fig. 12 Annotation in ARMARIUS

The system tracks the actions of the connected user; traces are stored with the actions and the affected objects (collections, pages, metadata…). User role is important in the tracing and the recommendation process; users have to be aware that their actions are traced. Therefore, the system shows the trace in construction to the user, enabling him to stop the tracing process whenever he wants. We introduce this fea-

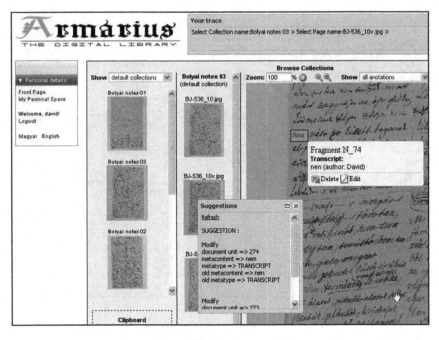

Fig. 13 ARMARIUS Recommendation window

ture in order to simplify the understanding of the recommendations, and to increase the confidence of users with a system that keeps track of their actions. Current user trace is exposed on the top of the window as shown in Fig. 13; it contains the action type, the name of the affected parameter by this action and its value. The system administrator is capable to activate/deactivate the tracing process of certain user(s) and delete user traces.

5.4 Recommendation System Based on CBR

Recommendations are generated by the intelligent assistant; the objective of the recommendations is to help users while they exploit the application to search or to annotate documents. The recommended actions are relative to the user last actions as well as his profile

The recommender system uses case-based reasoning to generate suggestions and recommends for example the user to see another document unit of his language, modify an annotation that was corrected by another user, and add new annotations to the document unit. Users can close the suggestion window if they are not interested in the assistance, or they can simply ignore the recommendations. An example of the suggestion window is shown in Fig. 13 where the system proposes the user to modify his annotation to "nem" according to a correction done on a similar fragment (274) in resembling situation.

6 Evaluation of CBR Use

The performance of the CBR system has been tested via the evaluation of the recommender system suggestions, the results showed that the use of CBR to generate relative recommendations to the user's work context is useful and help him in solving some problems in documents' annotation and search. Indeed, the challenge when assessing a recommender system is to select the appropriate metric. There is large diversity of metrics that have been used to evaluate the accuracy of recommended systems. The most used metrics are exploited to analyze the system recommendation quality or computation efficiency. Therefore, we applied two metrics to measure the *precision* of the proposed actions to users, and the *recall* of the system to produce sufficient recommendations.

Before starting the experimentation on our system, we chose a sample trace database to evaluate the recommendations, and then we conducted a series of experiments using real user-data from different user profiles. Furthermore, we were particularly interested to insure the effectiveness of episode similarity degree presented by our technique, and the high quality recommendations issued from this similarity of knowledge. Thus, in recommendation quality measurement we applied the metrics on the same database sample and we changed the action similarity degree to find the most appropriate threshold.

6.1 Metrics

The most exploited metrics to evaluate the system for its recommendation are *Precision* and *Recall*. In the quality evaluation, a recommender is evaluated based on whether its recommendations are relative to user's needs, this is Precision, and whether it generates enough recommendations, this is Recall. In our system, we use these two metrics to measure the ability of the recommender to return recommendations (possible next actions), from the set of all episodes that have the same level as the current user's episode, and to retrieve good quality suggestions.

Precision and recall are computed from Table 4. Suggested actions are separated into two classes: relevant or irrelevant. We also separated the actions in the trace database into suggested actions and not suggested actions.

Precision is defined as the ratio of relevant suggested actions to number of suggested actions, as shown in Eq. (13):

$$\text{Precision} = \frac{N_{RSA}}{N_{SA}} = \frac{N_{SA} - N_{ISA}}{N_{SA}} \qquad (13)$$

Table 4 Action categorizations in the system with respect to a given user situation

	Suggested Actions	Not Suggested Actions	Total
Relevant to user episode	N_{RSA}	N_{RNSA}	N_{RA}
Irrelevant to user episode	N_{ISA}	N_{INSA}	N_{IA}
Total	N_{SA}	N_{NSA}	

It represents the probability that a suggested action is relevant. Recall, shown in the Eq. (14), is defined as the ratio of relevant suggested actions to total number of relevant actions available. Thus, it represents the probability that a relevant action will be suggested.

$$\text{Recall} = \frac{N_{RSA}}{N_{RA}} = \frac{N_{SA} - N_{ISA}}{N_{SA} - N_{ISA} + N_{RNSA}} \tag{14}$$

One of the challenges to using precision and recall is that they must be considered together to evaluate completely the performance of an algorithm because they are inversely related. When more actions are returned, then the recall increases and precision decreases.

Precision and recall are inversely related and are dependent on the length of the result list returned to the user. When more actions are returned, then the recall increases and precision decreases. Moreover, several approaches have been taken to combine precision and recall into a single metric. One approach is the F1 metric [32] which combines precision and recall into a single number; it is one of the most popular techniques for combining these two metrics together in recommender system evaluation. It can be computed by the Eq. (15):

$$F1 = (2 * \text{Precision} * \text{Recall})/(\text{Precision} + \text{Recall}) \tag{15}$$

6.2 Dataset Used and Experimental Protocol

In our archive, we have different user groups (system administrators, librarians, historians, linguists, students). Currently, the system holds 10 original collections of handwritten manuscripts; of two categories: Hungarian manuscripts written by the mathematician "Bolyai" and Arabic manuscripts about Catechism. The experimental study was conducted on a test dataset of 133 manuscript pages, 706 registered traces of different users; traces are composed of 2226 episodes. To achieve the experiment, we asked 25 users who speak different languages and belong to different groups to accomplish a scenario. The tests were executed on users separately. Each time the data sets were reset to the initial values. When using recommendation systems, the user is the only person who can determine if a suggested action meets his requirements. All the users had the same dataset, they have been asked to repeat the same experience three times according to different percentages of similarity between actions. The three used thresholds are: 60, 70, and 90 %. The objective is to observe how the similarity threshold can affect the recommendations' quality and their quantity. During the test, we set the database to the initial dataset every time we changed the action similarity degree.

During the experimentation, users had to note down the performed actions, they had to judge the results of the system recommendations as relevant /irrelevant to their episode, and to note down the number N_{RSA}, N_{ISA} for every action they do. We extracted from the database all the similar episodes for each action to calculate the number of the actions that could be suggested and the number of the non-suggested actions N_{RNSA}, N_{INSA}. We used users' results and our calculations to measure the recall and the precision of the recommendations.

6.3 Experimental Results

The results of our experimentation to measure the precision and the recall of the system recommendations revealed that for a similarity action $= 60\%$, there were more actions to be recommended by the system, but the number of N_{ISA} was higher than the one when we used similarity action $= 70$ and 90 %. The experiments also exposed that with a similarity degree of 90 %, the users did not get recommendations for every action they made. We computed the precision and the recall from the entire user's experiments and our calculations; we preferred not to expose the experiments in the article because it comprises a huge table with user's actions as rows and the numbers of relevant/irrelevant, suggested/non-suggested actions in columns, the resulted precision and recall are illustrated in Fig. 14.

We notice that the recall is decreasing almost linearly when increasing the action similarity percentage. This is normal because increasing the similarity degree,

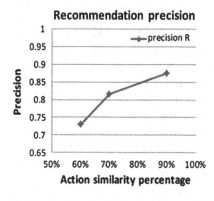
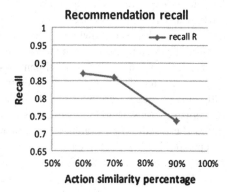

Fig. 14 Precision and Recall measurements for different action similarities thresholds

Fig. 15 Applying F1 metric
on the generated results from
precision and recall

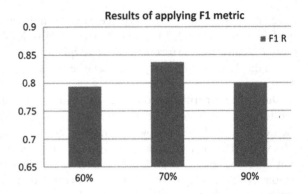

between the current user action and the registered actions, leads to fewer recommendations and thus to a smaller amount of recall.

As discussed previously, to measure the importance of the precision and recall metrics, we have to measure F1 metric. In (Fig. 15), we show the results of applying F1 metric on the results of precision and recall.

We notice that equilibrium results between precision and recall are obtained for an action similarity percentage of 70 %. Consequently, we fixed this similarity percentage in Algorithm1 to get similar episodes.

The disadvantage of using traces to generate recommendations is that it is relative to global users' interest. We remarked that collections that did not attract users have less chance to be recommended.

7 Conclusion

In this chapter we showed how user knowledge about handwritten manuscript collections is exploited by a case-based reasoning system, in a digital archive. The CBR is applied on users' traces, which are considered as users' experiences. We consider that the user knowledge might be represented either by the annotations he adds on the documents, or by the modifications he makes to annotations. Therefore, we proposed to integrate an intelligent system in the archive to trace the actions of authenticated users, and store traces in a relational database. To overcome the difficulty of making efficient search in the trace database, we chose to cut traces into episodes (cases) and register cases in a structured manner. Accordingly, users' traces (knowledge) are organized in a hierarchal structure where the basic unit is an episode; this later represents the entire work done on a document unit. Episodes could be simple or composed assembling small units into meaningful structures. The episodic trace structure in our system eases cases extraction and selection. Therefore, the proposed algorithms firstly compare current user episode with previous registered ones of other users to find similar situations. Similar cases (episodes) are obtained with a similarity greater than a predefined threshold. Secondly, the retrieved alike episodes are exploited by a recommender system to assist the user; this is done by adapting a solution for his situation while annotating and searching archive contents.

The presented approach is implemented as a prototype named ARMARIUS; the use of CBR to generate recommendations has been evaluated and has shown good results within the prototype. Our experiments confirmed the efficacy of using case-based reasoning in this domain. We believe that our archive model could also be used with other systems such as e-learning and e-commerce applications.

CV:

Reim DOUMAT received her Ph.D. in Computer Science at the Institute INSA-Lyon in 2010. Her research interests include multimedia information annotation and retrieval, Web digital archives, intelligent systems, user tracing systems, and Case-Based reasoning. She has published 8 journal articles and conference papers in these fields such as ACM DocEng, Springer ICCBR, Journal of Document numérique and journal of Multimedia Tools and Applications.

References

1. Digital Image Archive of Medieval Music: DIAMM. http://www.diamm.ac.uk/index.html. [Accessed 11-Nov-2009]
2. Avestan Digital Archive (ADA). http://ada.usal.es/. [Accessed 11-Nov-2009]
3. Rare Book and Manuscript Library of Columbia. http://www.columbia.edu/cu/lweb/indiv/rbml/. [Accessed 11-Nov-2009]
4. Gallica, bibliothèque numérique de la Bibliothèque nationale de France. http://gallica.bnf.fr/. [Accessed 25-Jan-2008]
5. INA, institut national de l'audiovisuel. http://www.ina.fr/. [Accessed: 25-Jan-2008]
6. Ramel, J.Y., Busson, S. Demonet, M.L.: AGORA: the Interactive Document Image Analysis Tool of the BVH Project. In: Proceedings of the Second International Conference on Document Image Analysis for Libraries (DIAL'06), pp. 145–155 (2006)
7. Kae, A., Learned-Miller, E.: Learning on the fly: font-Free approaches to difficult OCR problems. In: 10th International Conference on Document Analysis and Recognition, ICDAR '09, pp. 571–575. Barcelona (2009)
8. Kiamoto, A., Onishi, M., Ikezaki, T., Deuff, D., Meyer, E., Sato, S., Muramatsu, T., Kamida, R., Yamamoto, T., Ono, K.: Digital bleaching and content extraction for the digital archive of rare books. In: Proceedings of the Second international Conference on Document Image Analysis For Libraries (Dial'06), pp. 133–144. Washington, DC, USA (2006)
9. Dublin Core Metadata Element Set. http://dublincore.org/documents/dces/. [Accessed 12-Mai-2008]
10. Metadata Encoding and Transmission Standard (METS) Official Web Site. http://www.loc.gov/standards/mets/. [Accessed 13-Mai-2008]
11. TEI: Text Encoding Initiative. http://www.tei-c.org/index.xml. [Accessed 10-juin-2008]
12. MARC STANDARDS. http://www.loc.gov/marc/index.html. [Accessed 13-Mai-2008]
13. Digitization and Conversion:Mets/Alto: digital information accessible from press clipping to digital library. http://www.ccs-gmbh.com/alto/general.html. [Accessed 08-févr-2009]
14. Aleven, V.: Using background knowledge in case-based legal reasoning: A computational model and an intelligent learning environment. Artif. Intell. **150**(1–2), 183–237 (2003) (Elsevier)

15. Bichindaritz, I., Marling, C., et al.: Case-based reasoning in the health sciences: What's next? Artif. Intell. Med. **36**(2), 127–135 (2006) (Elsevier) (févr. 2006)
16. Ballesteros, M., Martin, R., Díaz-Agudo, B.: JADAWeb: A CBR System for Cooking Recipes. In: ICCBR Workshops Proceedings, pp. 179–188. Alessandria, Italiy (2010)
17. Blansché, A., Cojan, J., Dufour-Lussier, V., Lieber, J., Molli, P., Nauer, E., Skaf-Molli, H., Toussaint, Y.: TAAABLE 3: Adaptation of Ingredient Qualities and of Textual Preparations. In: ICCBR Workshops Proceedings, pp. 189–198. Alessandria, Italiy (2010)
18. Weber, B.G., Mateas, M.: Case-based reasoning for build order in real-time strategy games. In: Proceedings of the Fifth Artificial Intelligence and Interactive Digital Entertainment Conference, Stanford, California, USA (2009)
19. Uddin Ahmed, M., Begum, S., Olsson, E., Xiong, N., Funk, P.: Case-Based Reasoning for Medical and Industrial Decision Support Systems. In: Successful Case-based Reasoning Applications—I, SpringerLink., vol. 305/2010, pp. 7–52. Stefania Montani and Lakhmi Jain, (2012)
20. Pincho, N., Marques, V., Brito, A. Farinha, J.T.: E-learning by experience: how CBR can help. In: Proceedings of the 6th WSEAS International Conference on Distance Learning and Web Engineering, pp. 68–73 (2006)
21. Limthanmaphon, B., Zhang, Y.: Web service composition with case-based reasoning. In: Proceedings of the 14th Australasian database conference, Australia, vol. 17, p. 201–208 (2003)
22. Sun, Z., Han, J., Ma, D.: A Unified CBR Approach for Web Services Discovery, Composition and Recommendation. In: International Conference on Machine Learning and Computing, vol. 3, pp. 85–89 (2009)
23. Cordier, A. Mascret, B. Mille, A.: Dynamic Case-Based Reasoning for Contextual reuse of Experience. In: Workshop Proceedings of the Eighteenth International Conference on Case-Based Reasoning, Alessandria, Italiy, pp. 69–78 (2010)
24. Mille, A.: Traces Based Reasoning (TBR) Definition, illustration and echoes with story telling. LIRIS UMR 5205 CNRS/INSA de Lyon/Université Claude Bernard Lyon 1/Université Lumière Lyon 2/École Centrale de Lyon, Liris-2282 RR-LIRIS-2006-002, jan. 2006
25. Champin, P.-A., Prié, Y., Mille, A.: MUSETTE: Modelling USEs and Tasks for Tracing Experience. In: From structured cases to unstructured problem solving episodes—WS 5, Trondheim (NO), pp. 279–286 (2003)
26. Settouti, L.S., Prié, Y., Marty, J.-C., Mille, A.: A Trace-Based System for Technology-Enhanced Learning Systems Personalisation. In: Proceedings of the 2009 Ninth IEEE International Conference on Advanced Learning Technologies, pp. 93–97 (2009)
27. Cram, D., Jouvin, D., Mille, A.: Visualizing interaction traces to improve reflexivity in synchronous collaborative e-Learning activities. In: 6th European Conference on e-Learning, Copenhaguen, pp. 147–158 (2007)
28. Nathan, M., Harrison, C. Yarosh, S., Terveen, L., Stead, L., Amento, B.: CollaboraTV: making television viewing social again. In: 1st International Conference on Designing Interactive User Experiences for TV and Video, Silicon Valley, vol. 291, pp. 85–94. CA (2008)
29. Wang, Y., Stash, N., Aroyo, L., Hollink, L., Schreiber, G.: Using semantic relations for content-based recommender systems in cultural heritage. In: Proceedings of the Workshop on Ontology Patterns (WOP 2009), collocated with the 8th International Semantic Web Conference (ISWC-2009), vol. 516, Washington D.C., USA (2009)
30. Burke, R.: Hybrid recommender systems: survey and experiments. User Model. User-Adap. Inter. **12**(4), 331–370 (2002)
31. Adomavicius, G., Tuzhilin, A.: Towards the next generation of recommender systems: a survey of the state-of-the-art and possible extensions. IEEE Trans. Knowl. Data Eng. **17**(6), 734–749 (2005)
32. Cleverdon, C.W., Mills, J., Keen, M.: Factors Determining the Performance of Indexing Systems, vol. 2, Cranfield, UK (1966)

Chapter 7
TAAABLE: A Case-Based System for Personalized Cooking

Amélie Cordier, Valmi Dufour-Lussier, Jean Lieber, Emmanuel Nauer, Fadi Badra, Julien Cojan, Emmanuelle Gaillard, Laura Infante-Blanco, Pascal Molli, Amedeo Napoli and Hala Skaf-Molli

Abstract TAAABLE is a Case-Based Reasoning (CBR) system that uses a recipe book as a case base to answer cooking queries. TAAABLE participates in the Computer Cooking Contest since 2008. Its success is due, in particular, to a smart combination of various methods and techniques from knowledge-based systems: CBR, knowledge representation, knowledge acquisition and discovery, knowledge management, and natural language processing. In this chapter, we describe TAAABLE and its modules. We first present the CBR engine and features such as the retrieval process based on minimal generalization of a query and the different adaptation processes available. Next, we focus on the knowledge containers used by the system. We report on our experiences in building and managing these containers. The TAAABLE system has been operational for several years and is constantly evolving. To conclude, we discuss the future developments: the lessons that we learned and the possible extensions.

1 Introduction

TAAABLE is a Case-Based Reasoning system that provides cooking recipes in response to queries from users. When no recipe can be found that satisfies a query, an existing recipe is adapted. TAAABLE was developed to take part in the Computer

A. Cordier
LIRIS, CNRS, Université Claude Bernard Lyon 1, Villeurbanne Cedex, France

V. Dufour-Lussier · J. Lieber (✉) · E. Nauer · J. Cojan · E. Gaillard · L. Infante-Blanco · A. Napoli
LORIA, CNRS, INRIA, Université de Lorraine, Vandœuvre-lès-Nancy, France
e-mail: jean.lieber@loria.fr

P. Molli · H. Skaf-Molli
LINA, CNRS, Université de Nantes, Nantes Cedex 3, France

F. Badra
LIM&BIO, Université Paris 13, Bobigny Cedex, France

S. Montani and L. C. Jain (eds.), *Successful Case-based Reasoning Applications-2*,
Studies in Computational Intelligence 494, DOI: 10.1007/978-3-642-38736-4_7,
© Springer-Verlag Berlin Heidelberg 2014

Cooking Contest (CCC),[1] a competition held every year since 2008 during the International Conference on Case-Based Reasoning (ICCBR). The initial goal of the contest was to provide a common ground to allow for different CBR systems could compete and be compared. Therefore, it provided a benchmark for CBR tools. To this end, the contest organisers have made a *recipe book* available to participants and have created various challenges. Every year, each system is tested in front of a live audience at the conference and is evaluated by a panel of experts. The best system in each category is selected. TAAABLE has participated in the CCC since the first edition and has won several awards.

Several French researchers, from different laboratories, worked together to design and implement TAAABLE. They combined their skills and knowledge of various research issues: knowledge representation and knowledge management, case base organisation and representation, development of a similarity measure, adaptation knowledge acquisition, formal representation of preparations, retrieval and adaption strategies in CBR, etc. These different questions, investigated during the development of the TAAABLE project, are addressed in this chapter.

It is important to note that, although TAAABLE was primarily developed to enter the CCC and to address the CCC challenges, it also served as the subject of other research, which is also described in this chapter and published in various papers. TAAABLE has served as an application domain and as an experimentation and testing ground for three Ph.D. students. Furthermore, it was the starting point of a wider research project, funded by the French national agency for research: the Kolflow project.[2]

This chapter is organised as follows. In the rest of the introduction, we provide the reader with background knowledge on the CCC and we give a general overview of the TAAABLE system and architecture. In Sect. 2, we describe the inference engine of TAAABLE, which implements the fundamental operations of CBR: case retrieval and adaptation. These two operations are described in details. In Sect. 3, we study TAAABLE from the point of view of the knowledge it uses. We report the methods and tools we used in order to acquire, represent and manage this knowledge. We detail the knowledge containers used in the system: domain knowledge (ontologies), cases, and adaptation knowledge. Finally, we discuss the lessons we have learnt during the project and we briefly report our plans for future work.

The Computer Cooking Contest (CCC). The CCC is an annual competition that is organised by the international CBR community. Each year, the organisers provide the participants with a recipe book and define a set of challenges. Participants implement systems addressing these challenges by using the recipe book as a case base. Each participant describes their system in a technical paper and gives a live demonstration during the conference. Systems are evaluated against several *criteria* by a panel of scientists and cooking experts. The recipe book provided by the organisers is a simple XML file containing recipes. The schema of a recipe consists of one level

---

[1] http://computercookingcontest.net

[2] http://kolflow.univ-nantes.fr

```
<RECIPE>
<TI>Glutinous Rice with Mangoes</TI>
<IN>3 c Glutinous rice</IN>
<IN>1 1/2 c Coconut cream</IN>
<IN>1/2 c Sugar</IN>
<IN>1 ts Salt</IN>
<IN>1 1/4 c Coconut cream</IN>
<IN>2 tb Sugar</IN>
<IN>1/4 ts Salt</IN>
<IN>6 Ripe mangoes, well chilled</IN>
<IN>2 tb Sesame seeds, toasted</IN>
<PR>SEASONINGS SAUCE GARNISH Soak the rice in cold water for 2 hours.
Drain. Line a steamer with cheesecloth, heat steamer and lay rice on the
cheesecloth. Steam for 30 minutes or until cooked through. The rice will
become glossy. Mix the SEASONINGS ingredients in a large bowl and gently
mix in the hot steamed rice. Cover tightly and let soak for 30 minutes to
absorb the coconut flavour. Blend the SAUCE ingredients in a pot and heat
until it just reaches the boiling point. Let cool. Peel the mangoes, slice
lengthwise and remove the pits. Divide the rice among 6 plates. Place mango
slices on top and cover with the sauce. Sprinkle with the sesame seeds and
serve.
</PR>
</RECIPE>
```

Fig. 1 An example of recipe in the CCC recipe book

of tags only, with a title, a list of ingredients and a textual instructions that we call *preparation*. Figure 1 is an example of a recipe.

Using this recipe book is challenging because ingredients and preparations are written in natural language and may contain spelling mistakes. They cannot be used as the input of a computer program in their raw form. Therefore, a pre-processing step is required.

Over the years, the CCC organisers have proposed various challenges that are described hereafter. TAAABLE addressed all but the menu challenge.

The **main challenge**. "Given a query from a user, expressed in natural language, find a recipe satisfying this query. The proposed recipe must necessarily be adapted from a recipe of the recipe book." For example, "I would like a recipe with escarole endive and lemon juice but without onions" could be a query of the main challenge.

The **adaptation challenge**. "Given a recipe (from the recipe book) and a set of constraints, provide a suitable adaptation of this recipe." For example, given the recipe "Baked apple pancake" and the fact that I want to use bananas instead of apples, how do I adapt the recipe?

The **menu challenge**. "Given a query from a user, propose a suitable three-course menu." The user gives a list of ingredients and the goal is to retrieve three recipes using these ingredients.

The **healthy challenge**. This challenge is similar to the main challenge, but it includes additional constraints on special diets such as vegetarianism or a gluten-free diet.

The **open challenge**. This challenge was created in order to allow participants to investigate specific issues and demonstrate their results during the contest.

Major competing systems. From 2008 to 2010, many systems participated in the CCC. Even if they address the same problem, these systems are different in the way they index, retrieve and adapt cases, and in the type of adaptation they address. The "What's in the fridge?" system focuses on the recipe indexing using an active learning approach. The case retrieval and adaptation is based on a classical information retrieval technique, using Wordnet to compute similarity measures between the query and the set of recipes [24]. ColibriCook uses also a simple information retrieval technique: cases are retrieved using a similarity measure based on whether the elements of the query (ingredients, types, etc.) appear in the recipes [9]. Other systems like JadaCook [10] or CookIIS [15] take advantage of hierarchies to compute similarity. JadaCook uses a hierarchy of classes in order to compute the similarity between the query and the cases. The case retrieval takes into account the distance between the ingredients in the hierarchy. CookIIS also uses the hierarchy of ingredients, but do so in order to compute the similarity between ingredients. These similarities are then used both to compute the similarity between the query and the cases for case retrieval and to compute the adaptation. In addition, CookIIS uses ingredient similarities extracted from cooking websites [15]. CookIIS also addresses the adaptation of the recipe text, by making substitutions at the string level: a string (name of an ingredient) is replaced by another string (name of another ingredient). CookingCakeWF [22] also addresses the adaptation of the preparation, considering the instructions as a workflow. However, this system does not deal with the automatic transformation of the textual preparation into a workflow.

An overview of the TAAABLE user interface. The user interface of TAAABLE[3] consists of a query and result interfaces (see Fig. 2). The upper part of the figure shows a query input (dessert dish, with rice and figs, no particular diet). The lower part of the figure shows the answer to the query–only one recipe is returned in this example–and the associated adaptation (replace Mango with Fig).

The user can see the original recipe as well as the details of the adaptation by clicking on the available links. Figure 3 shows the result of the adaptation of the recipe "Glutinous Rice with Mangoes" obtained by replacing mangoes with figs. The adaptation is composed of three parts. First, the ingredient substitutions are mentioned. Second, the ingredient quantities are adapted. Third, the textual preparation of the recipe is adapted. These three types of adaptation and the processes that are involved are detailed in Sect. 2.

An overview of the TAAABLE architecture. As a CBR system, TAAABLE takes as input a query from a user and searches for a similar case in the case base, adapts the retrieved case, and presents the result. It uses most notably a case base (the recipe book), a domain knowledge base (about cooking), and an adaptation knowledge base. Early version of the system used knowledge bases stored in XML files and built and managed by hand, but this proved to be impossible to maintain. For this reason, we

---

[3] http://taaable.fr

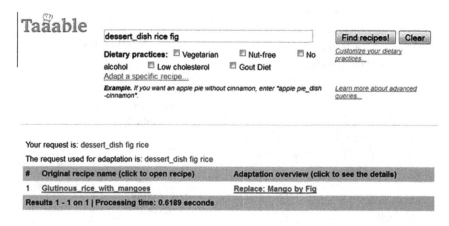

Fig. 2   An illustration of the TAAABLE user interface

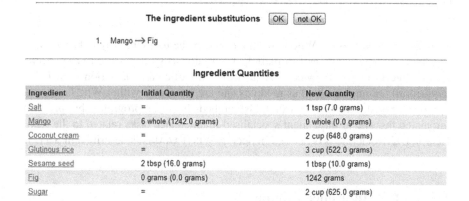

Fig. 3   An example of adaptation results

decided to use a common tool to manage all the knowledge sources: a semantic wiki named WIKITAAABLE.

Figure 4 presents the general architecture of TAAABLE. The TAAABLE user interface makes it possible to query the system and to display the results provided by the TAAABLE CBR engine. The CBR engine is connected to WIKITAAABLE, which con-

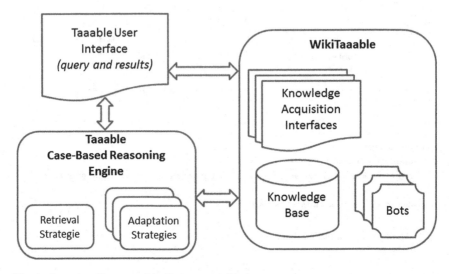

Fig. 4 General architecture of the TAAABLE application

tains the knowledge base. WIKITAAABLE provides the users the way to browse and edit the knowledge units and to participate in knowledge acquisition tasks. Some specific interfaces for knowledge acquisition (in particular, adaptation knowledge acquisition) are implemented as modules and are integrated within WIKITAAABLE. It must be noted that WIKITAAABLE also embeds bots (i.e. programs that perform automated tasks) that implement specialised maintenance tasks related to the wiki. TAAABLE CBR engine is described in Sect. 2. WIKITAAABLEis described in Sect. 3.

2 TAAABLE Inference Engines

2.1 Case Retrieval

In TAAABLE, the retrieval of recipes consists in selecting recipes R from the recipe base `Recipes` (used as a case base[4]), where R is a best match to the query Q. This best match is computed by finding a generalisation $\Gamma(Q)$ of Q that is minimal (according to a cost function) and such that there is at least one recipe exactly matched by $\Gamma(Q)$. This matching takes into account the domain knowledge DK.

The domain knowledge DK can be considered as a set of axioms $a \Rightarrow b$ in propositional logic, where a and b are propositional variables representing recipe classes. For example, `lemon` (resp., `citrusfruit`) represents the class of recipes having lemons (resp., citrus fruits) as ingredients and the axiom `lemon` $\Rightarrow$ `citrusfruit`

---

[4] For the adaptation challenge, the recipe base contains a sole case, the recipe that must be adapted.

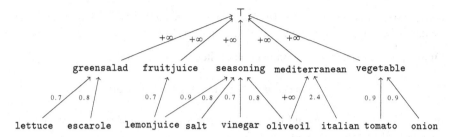

Fig. 5 An excerpt of domain knowledge with costs. The edge $a \xrightarrow{c} b$ means that $a \Rightarrow b \in$ DK and $\text{cost}(a \rightsquigarrow b) = c$

states that every recipe with lemon is a recipe with citrus fruit. In fact, each food name x is interpreted as "the class of recipes having x as ingredient". There are other propositional variables, such as `mediterranean` that represents the class of Mediterranean recipes. Therefore, `oliveoil` $\Rightarrow$ `mediterranean` states that every recipe with olive oil is a Mediterranean recipe. From an implementation viewpoint, DK is a hierarchy, i.e. a directed acyclic graph whose nodes are propositional variables and edges $a \longrightarrow b$ correspond to axioms $a \Rightarrow b$. $\top$, the root of this hierarchy, denotes the recipe universe.

Section 3.1 presents the acquisition of a domain ontology $\mathcal{O}$ given by class hierarchies, organised with a "more specific than" relation denoted by . For instance, there is a food hierarchy in $\mathcal{O}$ stating, in particular, that LemonCitrusFruit. The domain knowledge DK used for case retrieval and case adaptation is based on this ontology with a change of formalism (from the ontology language to propositional logic). In particular, `lemon` represents the recipes with lemon and, since `Lemon` is a subclass of `CitrusFruit`, each recipe with lemon is a recipe with fruit, hence the axiom `lemon` $\Rightarrow$ `citrusfruit` of DK.

Figure 5 is an excerpt of the domain knowledge hierarchy (the valuation on the edges represents costs that are explained further).

The recipe base `Recipes` is the case base of the system. These recipes are provided by the CCC organisers (there were about 900 source recipes in the first CCC and about 1500 ones for the further CCCs). A recipe R $\in$ `Recipes` is described by a shallow XML document that has to be transformed within a representation formalism. For case retrieval, R is transformed into $idx(R)$ (the index of R which is a conjunction of literals) thanks to an annotation process presented in Sect. 3.4. For example,

$$idx(R) = \texttt{lettuce} \wedge \texttt{vinegar} \wedge \texttt{oliveoil} \wedge \texttt{tomato} \wedge \textit{nothing else} \quad (1)$$

is a formal and abstracted representation of the recipe R whose ingredients are a lettuce, vinegar, olive oil, tomatoes, and nothing else. A closed world assumption is associated to $idx(R)$: if a property cannot be deduced from $idx(R)$ and the domain knowledge DK, then it is considered as false. In other words, if $idx(R) \not\models_{\text{DK}} a$ then the term *nothing else* is a conjunction of literals containing the literal $\neg a$. In the example above, $idx(R) \models_{DK} \neg$meat $\wedge \neg$fish. It can be noticed that $idx(R)$ has exactly one model: for each propositional variable a, either $idx(R) \models_{DK} a$ or $idx(R) \models_{DK} \neg a$.

Query Q is represented by a conjunction of literals. For example,

$$Q = \text{escarole} \wedge \text{lemonjuice} \wedge \neg\text{onion} \tag{2}$$

represents the query "I would like a recipe with escarole and lemon juice (among other ingredients) but without onions."

Recipe retrieval is described in the Algorithm 1 and is an application of smooth classification as described in [19]. It consists in finding a generalisation $\Gamma(Q)$ of Q that exactly matches at least one recipe of the recipe base (i.e. there exists R ∈ Recipes such that $idx(R) \models_{DK} \Gamma(Q)$). Lines 4 and 5 require more explanation.

The generalisation $\Gamma(Q)$ of Q is searched in a state space where each state is a conjunction of literals, the initial state is Q and the successors of a state s is the set of the states $\gamma(s)$ where γ is a state transition that is either

---

Retrieval (Q, Recipes, DK) : ($\{R_i\}_i$, Γ)

Input   a query Q, the recipe base Recipes, the domain knowledge DK
Output   a set of recipes $\{R_i\}_i \subseteq$ Recipes and a generalisation function Γ such that each R_i
          exactly matches $\Gamma(Q)$

1: $\Gamma \leftarrow$ identity function
2: $\{R_i\}_i \leftarrow \emptyset$
3: **while** $\{R_i\}_i = \emptyset$ **do**
4:     $\Gamma \leftarrow$ next generalisation
5:     $\{R_i\}_i \leftarrow \{R_i \in$ Recipes $| idx(R_i) \models_{DK} \Gamma(Q)\}$
6: **end while**

algorithm 1: TAAABLE retrieval algorithm.

- a one step generalisation of a positive literal in s according to the ontology, e.g. if s contains the literal lemon, then $\gamma =$ lemon $\rightsquigarrow$ citrusfruit ($x \rightsquigarrow y$ is the substitution of x with y) is one of the γ's if lemon $\Rightarrow$ citrusfruit is an axiom of the ontology; or
- a removal of a negative literal of s, e.g. if s contains $\neg$garlic, then $\gamma = \neg$garlic $\rightsquigarrow \top$ is one of the γ's.

Thus, each generalisation of Q is a $\Gamma(Q)$ where Γ is a composition of γ's. This state space is searched according to an increasing value of cost(Γ) where cost is an additive cost function (cost($\Gamma_2 \circ \Gamma_1$) = cost(Γ_1) + cost(Γ_2)) such that the cost of the identity function is null and cost(γ) > 0 for each state transition γ. So, the next generalisation mentioned in line 4 is the generalisation of cost immediately greater than the cost of Γ. Technically, this next generalisation is the first element of a list of generalisations ordered by an increasing cost, as in a classical A* search [23].

Line 5 of the algorithm is implemented in TAAABLE using a hierarchical classification of $\Gamma(Q)$ in a hierarchy containing the recipe indexes $idx(R)$. This mechanism is a classical deductive inference described, e.g. in [1].

The cost function has to be defined for each $a \rightsquigarrow b$ such that $a \Rightarrow b \in$ DK. Indeed, every generalisation Γ is a composition of n substitutions $a_i \rightsquigarrow b_i$, and $\text{cost}(\Gamma) = \sum_{i=1}^{n} \text{cost}(a_i \rightsquigarrow b_i)$ (since cost is additive). It is assumed that the recipe base, Recipes, constitutes a homogeneous sampling of recipes (in the recipe space) and thus, that $\mu(x) = \dfrac{\mathcal{N}(x)}{\mathcal{N}(\top)}$ is a good approximation of the proportion of recipes of class x, where $\mathcal{N}(x)$ is the number of recipes of the class x:

$$\mathcal{N}(x) = \left| \{R \in \text{Recipes} \mid idx(R) \models_{\text{DK}} x\} \right|$$

Thus, if $a \Rightarrow b \in$ DK, $\mu(b) - \mu(a)$ is the proportion of recipes of class b that are not recipes of class a and thus is characteristic of the risk made by the generalisation $a \rightsquigarrow b$. Therefore, a first idea is to define $\text{cost}(a \rightsquigarrow b) = \mu(b) - \mu(a)$. Now, a more accurate definition of cost is based on the adaptation-guided retrieval principle [25] stating that adaptation knowledge should be taken into account during retrieval. In particular, it is worth noticing that if a and b are ingredient-based recipe classes (e.g. $a = $ lemon and $b = $ citrusfruit) then the adaptation will be substitution-based whereas if a and b are, for instance, location-based recipe classes (e.g. $a = $ italian and $b = $ mediterranean) then the adaptation will be a mere copy of the source recipe, involving more risks. Therefore, the adaptation cost depends on the *type* τ of propositional variables a and b. This type is associated to each propositional variable, e.g. the type of lemon and of citrusfruit is $\tau = $ ingredient and the type of italian and of mediterranean is $\tau = $ location. Thus, a coefficient K_τ depending on the type τ of classes a and b is used and the cost is defined by:

$$\text{cost}(a \rightsquigarrow b) = K_\tau \frac{\mathcal{N}(b) - \mathcal{N}(a)}{\mathcal{N}(\top)} \quad \text{with } \tau \text{ the type of } a \text{ and } b$$

(if a and b are of different types, the generalisation is forbidden, e.g. cost (oliveoil $\rightsquigarrow$ mediterranean) $= +\infty$). Since the adaptation related to ingredients is assumed to be less risky than the one related to location, $K_{\text{ingredient}}$ is much lower than K_{location}. In practice, the following values give satisfying results: $K_{\text{ingredient}} = 1$ and $K_{\text{location}} = 10$.

For example, let us consider the domain knowledge DK with the costs given by Fig. 5, the query of Eq. (2) and a case base containing only one case: the recipe R indexed by the $idx(R)$ of Eq. (1). Then, the retrieval process generates the following generalisation functions Γ_t (the composition operator is denoted by o):

$\Gamma_0 = \text{identity function}$ (cost 0)
$\Gamma_1 = \text{lemonjuice} \rightsquigarrow \text{fruitjuice}$ (cost 0.7)
$\Gamma_2 = \text{escarole} \rightsquigarrow \text{greensalad}$ (cost 0.8)
$\Gamma_3 = \text{lemonjuice} \rightsquigarrow \text{seasoning}$ (cost 0.9)
$\Gamma_4 = \text{lemonjuice} \rightsquigarrow \text{fruitjuice} \circ \text{escarole} \rightsquigarrow \text{greensalad}$ (cost 1.5)
$\Gamma_5 = \text{lemonjuice} \rightsquigarrow \text{seasoning} \circ \text{escarole} \rightsquigarrow \text{greensalad}$ (cost 1.7)

So, the result of the retrieval process is $(\{R\}, \Gamma)$ with $\Gamma = \Gamma_5$ since $idx(R) \models_{DK} \Gamma_5(Q)$ and $idx(R) \not\models_{DK} \Gamma_t(Q)$ for $t < 5$.

2.2 Case Adaptation

Let $\{R_i\}_i$ be the set of retrieved cases. Given a retrieved recipe $R \in \{R_i\}_i$ and the query Q, adaptation aims at pointing out modifications of R so that it answers the query Q. Several adaptation processes have been implemented in TAAABLE. The first (Sect. 2.2.1) uses only the result of the retrieval: R and Γ. The second (Sect. 2.2.2) uses adaptation knowledge in the form of ingredient adaptation rules. These two adaptation processes just give substitutions of ingredient types, regardless of ingredient quantities. By contrast, the third adaptation process (Sect. 2.2.3) addresses the issue of ingredient quantities. Finally, Sect. 2.2.4 addresses the issue of textual adaptation of the preparation of R.

2.2.1 Basic Adaptation

The basic adaptation procedure uses only R, Γ and Q and no other piece of knowledge. It is based on the following sequence of relations that are satisfied since $R \in \{R_i\}_i$ and $(\{R_i\}_i, \Gamma)$ is the result of the retrieval process:

$$idx(R) \models_{DK} \Gamma(Q) \xleftarrow{\Gamma} Q \tag{3}$$

This sequence of relations involves a matching between the recipe ingredients and the positive literals of Q. For the current example, this matching is based on the following composition:

$$\begin{aligned} \text{lettuce} \rightsquigarrow \text{greensalad} \circ \text{vinegar} \rightsquigarrow \text{seasoning} \\ \circ \text{greensalad} \rightsquigarrow \text{escarole} \circ \text{seasoning} \rightsquigarrow \text{lemonjuice} \end{aligned} \tag{4}$$

The first line is related to $\models_{DK}$. The second line is related to $\xleftarrow{\Gamma}$, i.e., to $\xmapsto{\Gamma^{-1}}$. This composition can be simplified (using the equation $a \rightsquigarrow b \circ b \rightsquigarrow c = a \rightsquigarrow c$ and licit permutations in the composition) into:

$$\texttt{lettuce} \rightsquigarrow \texttt{escarole} \circ \texttt{vinegar} \rightsquigarrow \texttt{lemonjuice} \qquad (5)$$

which describes which substitutions can be made on the recipe R to answer the query Q.

More generally, basic adaptation consists in building a composition of substitutions as in (4) and then to simplify it (when it is possible) to obtain a substitution that is applied on the retrieved recipe R. Technically, the elements of the substitutions are pointed out in two steps:

1. Following $idx(\text{R}) \models_{\text{DK}} \Gamma(\text{Q})$, generalisation substitutions are built. Indeed, it can be shown, for φ and χ, conjunctions of literals, that if $\varphi \models_{\text{DK}} \chi$ then there exists a composition G of substitutions of the form $a \rightsquigarrow b$ where $a \Rightarrow b \in \text{DK}$ such that $G(\varphi)$ is equivalent to χ (given the domain knowledge DK).

2. Following $\Gamma(\text{Q}) \overset{\Gamma}{\longleftarrow} \text{Q}$, specialisation substitutions γ^{-1} are built: if $\Gamma = \gamma_q \circ \ldots \circ \gamma_1$ then $\Gamma^{-1} = \gamma_1^{-1} \circ \ldots \circ \gamma_q^{-1}$. A generalisation $\gamma = a \rightsquigarrow b$ will lead to the specialisation $\gamma^{-1} = b \rightsquigarrow a$. A generalisation $\gamma = \neg a \rightsquigarrow \top$ will lead to the specialisation $\gamma^{-1} = \top \rightsquigarrow \neg a$, i.e., to the removal of a in the recipe.

2.2.2 Rule-Based Adaptation

The TAAABLE adaptation engine can also take advantage of adaptation knowledge in the form of adaptation rules. Such a rule states that in a given context $\mathcal{C}$, some ingredients $\mathcal{F}$ can be replaced by other ingredients $\mathcal{B}$ ($\mathcal{C}$, $\mathcal{F}$ and $\mathcal{B}$ are the contexts, the "from part" and the "by part" of the adaptation rule). For example, let us consider the following piece of knowledge:

In a recipe with green salad,

vinegar can be replaced by lemon juice and salt.

This piece of knowledge can be represented by an adaptation rule with $\mathcal{C} = \texttt{greensalad}$, $\mathcal{F} = \texttt{vinegar}$ and $\mathcal{B} = \texttt{lemonjuice} \wedge \texttt{salt}$. Such an adaptation rule can be encoded by a substitution $\sigma = \mathcal{C} \wedge \mathcal{F} \rightsquigarrow \mathcal{C} \wedge \mathcal{B}$. In the example:

$$\texttt{greensalad} \wedge \texttt{vinegar} \rightsquigarrow \texttt{greensalad} \wedge \texttt{lemonjuice} \wedge \texttt{salt} \qquad (6)$$

Let AK denote the adaptation knowledge, i.e., the finite set of substitutions σ representing adaptation rules. A cost associated to each $\sigma \in \text{AK}$, $\texttt{cost}(\sigma) > 0$, is also assumed to be known. Given AK, the domain knowledge DK, a query Q, and a retrieved recipe R, rule-based adaptation combines the use of adaptation rules and of the generalisation of the query according to DK. It aims at building a sequence of relations of the form

$$idx(\text{R}) \overset{\Sigma}{\longmapsto} \Sigma(idx(\text{R})) \models_{\text{DK}} \Lambda(\text{Q}) \overset{\Lambda}{\longleftarrow} \text{Q} \qquad (7)$$

where Σ is a (possibly empty) composition of adaptation rules $\sigma \in \text{AK}$, Λ is a generalisation function (that may be different from the generalisation function Γ returned by the retrieval process) and $\text{cost}(\Sigma) + \text{cost}(\Lambda)$ is minimal.

The sequence (7) is called an *adaptation path*. The adaptation consists in following this path: first, the composition Σ of adaptation rules is applied on the recipe R, then generalisations corresponding to $\Sigma(idx(\text{R})) \models_{\text{DK}} \Lambda(\text{Q})$ are applied, and finally, specialisations corresponding to $\Lambda(\text{Q}) \overset{\Lambda}{\longleftarrow} \text{Q}$ are applied. It can be noted that the second and third steps correspond to the adaptation path (7) when $\Sigma = $ identity function and $\Lambda = \Gamma$.

Therefore, the main algorithmic difficulty of rule-based adaptation is to build an adaptation path. Once again, the technique used is based on a best-first search in a state space. For this search, a state is an ordered pair (Σ, Λ), the initial state corresponds to $\Sigma = \Lambda = $ identity function, (Σ, Λ) is a final state if (7) is satisfied, the cost associated to a state (Σ, Λ) is $\text{cost}(\Sigma) + \text{cost}(\Lambda)$ and the successors of a state (Σ, Λ) are the states (Σ', Λ) and the states (Σ, Λ') such that

- The Σ' substitutions are such that $\Sigma' = \sigma \circ \Sigma$ with $\sigma \in \text{AK}$ and σ is applicable on $\Sigma(idx(\text{R}))$;
- The Λ' substitutions are such that $\Lambda' = \gamma \circ \Lambda$ with γ, a generalisation based on DK that is applicable on $\Lambda(\text{Q})$ (γ has either the form $a \rightsquigarrow b$ or the form $\neg a \rightsquigarrow \top$).

It can be noticed that the search space contains at least one final state: in particular (Σ, Λ) with $\Sigma = $ identity function and $\Lambda = \Gamma$ satisfies (7), since (3) is satisfied (R being a retrieved case with a generalised query $\Gamma(\text{Q})$), thus the algorithm terminates. Moreover, this shows that rule-based adaptation amounts to basic adaptation when there is no available adaptation rule ($\text{AK} = \emptyset$).

For example, with $idx(\text{R})$ and Q defined by (1) and (2), with the domain knowledge with costs of Fig. 5, and with $\text{AK} = \{\sigma\}$, σ being the adaptation of (6) and $\text{cost}(\sigma) = 0.2$, then the adaptation gives $\Sigma = \sigma$, $\Lambda = $ escarole $\rightsquigarrow$ greensalad with a cost $0.2 + 0.8 = 1$ which involves an adaptation of the recipe based on the substitution

$$\text{lettuce} \rightsquigarrow \text{escarole} \circ \text{vinegar} \rightsquigarrow \text{lemonjuice} \wedge \text{salt}$$

Taking into account specific adaptation knowledge. For some recipes R, there exist specific adaptation rules, applicable in the context of the recipes ($\mathcal{C} = \text{R}$). The set of the specific adaptation rules associated with R is denoted by AK_R. The provenance of these rules is detailed in Sect. 3.5. This occurs in particular when there are variants in the original text of the recipe; if butter is an ingredient of R that is mentioned to be replaceable by margarine, then $\sigma = \text{butter} \rightsquigarrow \text{margarine} \in \text{AK}_\text{R}$ and, since this variant is mentioned in the recipe, the cost of this rule is assumed to be 0.

Taking into account these specific adaptation rules consists simply in considering AK_R, where R is the retrieved recipe, in addition to the adaptation rules of AK: adaptation of R is based on $\text{AK} \cup \text{AK}_\text{R}$.

2.2.3 Adaptation of the Ingredient Quantities

The basic adaptation of Sect. 2.2.1 points out a composition of ingredient substitutions (like Eq. (5)) and works at the abstract representation level of the recipe index $idx(R)$ of the retrieved recipe R. This section and the next one are interested in more concrete elements of R; here, adaptation deals with the adaptation of quantities and reuse the result of the basic adaptation, i.e., the composition of ingredient substitutions $a_i \rightsquigarrow b_i$.

The ingredient quantities are adapted following the revision-based adaptation principle. This principle is detailed in [5]. Roughly said, a revision-based adaptation consists in modifying the source case minimally so that it is consistent with the query, while taking into account the domain knowledge. Such minimal modification can be performed by a belief revision operator and is frequently modelled thanks to a distance.

This adaptation of ingredient quantities is performed after the retrieval of a source recipe, and a first adaptation, as a conjunction of (Boolean) ingredient substitutions. Each substitution replaces an ingredient a from the source recipe with another ingredient b. So the quantity adaptation takes into account the following inputs:

- The formula $idx(R)$ in propositional logic that represents the source recipe (see, e.g., Eq. (1)). Dish types and origins are ignored here. In this section, the following example, corresponding to the recipe "Baked apple pancake", is considered:

$$idx(R) = \texttt{flour} \wedge \texttt{milk} \wedge \texttt{granulatedsugar} \wedge \texttt{egg} \wedge \texttt{apple} \wedge \textit{nothing else}$$

- The formula Q in propositional logic that represents the query made by the user. In this section, the following example is considered:

$$Q = \texttt{banana} \wedge \neg\texttt{chocolate}$$

- A composition of ingredient substitutions $a_i \rightsquigarrow b_i$. In the example of this section, there is only one substitution $a_1 \rightsquigarrow b_1$ with:

$$a_1 = \texttt{apple} \qquad\qquad b_1 = \texttt{banana}$$

Representation formalism. The formalism used is based on numerical variables and linear constraints. Each variable corresponds to the quantity of an ingredient type. The list of ingredient types is reduced to those appearing in the source recipe, in the query, or in the substitutions. This list of ingredient types is closed by deduction w.r.t. domain knowledge: all the ingredient types that generalise one of the element of this list is also included. The following ingredients are considered in the example: `flour`, `cerealproduct`, `milk`, `dairyproduct`, `egg`, `granulatedsugar`, `chocolate`, `seasoning`, `apple`, `pomefruit`, `banana`, `fruit`, `ingredient`.

The ingredient amounts given in recipes are expressed in different units (grams, millilitres, teaspoons, pieces, ...). According to the unit used, the variables take their value in $\mathbb{R}_+$ (grams, millilitres) or $\mathbb{N}$ (spoons, pieces). To express constraints between the ingredient quantities, the quantities are converted into grams. To do this, we introduce, if needed, an additional variable per ingredient that represents its amount in grams. The quantity of an ingredient can be represented by several variables, conversions between the values of these variables are given by domain knowledge.

In addition to ingredient quantities, nutritional values are also taken into account. The list of these nutritional values is given in the food pages (see Fig. 6). A variable is introduced for each nutritional value. The quantities that have a significant impact on the taste and appearance of the final recipe are the amount of sugars, fat and water. They are given more importance in the adaptation calculus than the other variables. The other quantities have, however, a dietary interest and we could consider these values for requests over specific diets.

Some nutritional variables cannot be converted into grams (for example, the energy contained in the ingredients is in calories), so we consider only one variable per nutritional value with the unit in which it is expressed (see Fig. 6).

Other values could be taken into account for the quantity adaptation, for instance the bitterness of ingredients or values that express their taste strength. But we lack the knowledge to compute them.

The recipes are then represented with variables $x_1, \ldots, x_n$ that respectively take their values in value spaces $\mathcal{U}_1, \ldots, \mathcal{U}_n$ with either $\mathcal{U}_i = \mathbb{R}_+$ or $\mathcal{U}_i = \mathbb{N}$. So a recipe is represented by a value in $\mathcal{U} = \mathcal{U}_1 \times \cdots \times \mathcal{U}_n$. In the example, the same notation for the numerical variables as for the propositional variables is used, with units in underscore. For instance, the variable corresponding to the amount of apples in grams is written $\mathtt{apple_g}$, the variable corresponding to the amount of apples in units (pieces) is written $\mathtt{apple_u}$. As presented further, the cooking knowledge used can be expressed as linear constraints over variable values.

Representation of recipes. The recipe R is represented by a conjunction $clc(\mathrm{R})$ of linear constraints. For each ingredient of the recipe, a linear constraint $(x = v)$ is added to $clc(\mathrm{R})$, where x is the variable corresponding to this ingredient and v is its value in the recipe. For any ingredient type p not listed in the recipe, i.e. such that $idx(\mathrm{R}) \models_{\mathrm{DK}} \neg p$, the formula $(p_g = 0)$ is added to $clc(\mathrm{R})$. For example, the recipe "Baked apple pancake" is represented by

$$clc(\mathrm{R}) = (\mathtt{flour_{cup}} = 1) \wedge (\mathtt{milk_{cup}} = 1) \wedge (\mathtt{egg_u} = 4)$$
$$\wedge (\mathtt{granulatedsugar_{cup}} = 1) \wedge (\mathtt{apple_u} = 2) \wedge (\mathtt{banana_g} = 0)$$

Representation of the domain knowledge. The linear constraints related to the domain knowledge used for this adaptation are denoted by $clc(\mathrm{DK})$ and correspond to unit conversions, the food hierarchy given in the domain knowledge DK, and the nutritional data for each food.

Unit conversion. For every ingredient quantity expressed in two units, with a variable $\mathtt{food_{[unit]}}$ that represents its amount in the unit [unit] and a variable $\mathtt{food_g}$ that represents its amount in grams, the following formula is added to $clc(\mathrm{DK})$:

Nutritional values	
Nutritional value per 100 g (3.5 oz)	
Energy	52 kcal (220 kJ)
Carbohydrates	13.81 g
Sugars	10.39 g
Dietary fiber	2.4 g
Fat	0.17 g
Protein	0.26 g
Water	85.56 g
Vitamin A (equiv.)	3 µg (0%)
Thiamine (Vit. B1)	0.017 mg (1%)
Riboflavin (Vit. B2)	0.026 mg (2%)
Niacin (Vit. B3)	0.091 mg (1%)
Pantothenic acid (Vit. B5)	0.061 mg (1%)
Vitamin B6	0.041 mg (3%)
Folate (Vit. B9)	3 µg (1%)
Vitamin C	4.6 mg (8%)
Calcium	6 mg (1%)
Iron	0.12 mg (1%)
Magnesium	5 mg (1%)
Phosphorus	11 mg (2%)
Potassium	107 mg (2%)
Sodium	1 mg (0%)
Zinc	0.04 mg (0%)
Percentages are relative to US recommendations for adults. Source: USDA Nutrient database Corresponding ingredient: APPLES,RAW,WITH SKIN	

Nutritional values	
Nutritional value per 100 g (3.5 oz)	
Energy	89 kcal (370 kJ)
Carbohydrates	22.84 g
Sugars	12.23 g
Dietary fiber	2.6 g
Fat	0.33 g
Protein	1.09 g
Water	74.91 g
Vitamin A (equiv.)	3 µg (0%)
Thiamine (Vit. B1)	0.031 mg (2%)
Riboflavin (Vit. B2)	0.073 mg (5%)
Niacin (Vit. B3)	0.665 mg (4%)
Pantothenic acid (Vit. B5)	0.334 mg (7%)
Vitamin B6	0.367 mg (28%)
Folate (Vit. B9)	20 µg (5%)
Vitamin C	8.7 mg (15%)
Calcium	5 mg (1%)
Iron	0.26 mg (2%)
Magnesium	27 mg (7%)
Phosphorus	22 mg (3%)
Potassium	358 mg (8%)
Sodium	1 mg (0%)
Zinc	0.15 mg (2%)
Percentages are relative to US recommendations for adults. Source: USDA Nutrient database Corresponding ingredient: BANANAS,RAW	

Fig. 6 Nutritional values extracted from WIKITAAABLE pages *Apple* and *Banana*. This data comes from a copyright-free database of the *United States Department of Agriculture* (USDA, http://www.ars.usda.gov/)

$$(\text{food}_g = \alpha \cdot \text{food}_{[\text{unit}]})$$

where α is the mass in grams for 1 [unit] of food.

Remark 1 We use the word "conversion" with a broader meaning than in physics since we also perform the conversion between values expressed in heterogeneous units. For instance, 1 cup of flour weighs 250 g. For ingredients measured in pieces— like eggs, bananas and apples in the example—we consider the typical mass of a piece, these values are also taken from the USDA database (see Fig. 6). When the corresponding values are available, qualifiers given to the ingredients are taken into account. For instance, the recipe "Baked Apple Pancake" contains large apples, that weigh 223 g each according to USDA data, a small apple only weighs 101 g according to the same dataset.

The following unit conversions are used in the example:

$$(\text{apple}_g = 223 \cdot \text{apple}_u) \qquad \text{A (large) apple weighs 223 g.}$$
$$(\text{flour}_g = 120 \cdot \text{flour}_{\text{cup}}) \qquad \text{A cup of flour weighs 120 g.}$$

Food hierarchy. The food hierarchy is expressed by linear constraints over variable values of foods related by a generalisation relation. The quantity of a general food in a recipe is equal to the sum of the quantities of the more specific foods used in the recipe.

So, if a food GenFood covers the specific foods food1, ..., foodk, the following formula is added to $clc(DK)$:

$$(\texttt{GenFood}_g = \texttt{food1}_g + \cdots + \cdots \texttt{foodk}_g)$$

Remark 2 By "specific foods", we mean ingredient types for which there are no more specific food in the ingredient list. In the example, the value of $\texttt{fruit}_g$ is given by the following equality:

$$\texttt{fruit}_g = \texttt{apple}_g + \texttt{banana}_g$$

The term $\texttt{pomefruit}_g$ does not appear in this sum since we already have $\texttt{pomefruit}_g = \texttt{apple}_g$ as pomefruit generalises apple: the mass of apples in $\texttt{fruit}_g$ must not be counted twice.

We decided to take the equalities $\begin{cases} \texttt{fruit}_g = \texttt{apple}_g + \texttt{banana}_g \\ \texttt{pomefruit}_g = \texttt{apple}_g \end{cases}$

rather than $\begin{cases} \texttt{fruit}_g = \texttt{pomefruit}_g + \texttt{banana}_g \\ \texttt{pomefruit}_g = \texttt{apple}_g \end{cases}$

to avoid the problems raised by multiple specialisation (recall that the hierarchy is not a tree). For instance, broccoli direct superclasses are inflorescentvege table and cabbage,[5] so if we write the following equalities:

$$\texttt{vegetable}_g = \texttt{inflorescentvegetable}_g + \texttt{cabbage}_g$$
$$\texttt{inflorescentvegetable}_g = \texttt{broccoli}_g$$
$$\texttt{cabbage}_g = \texttt{broccoli}_g$$

broccoli mass would be counted twice in the vegetable mass. Instead of the first equality, we write $\texttt{vegetable}_g = \texttt{broccoli}_g$.

Remark 3 An additional problem arises for some recipes when the quantities of foods food1 and food2 are given in the recipe, although food1 is classified as more general than food2. This generates inconsistencies between the constraints. For example, the recipe "*Vegetables in crock pot*" contains 1 cup of broccoli (91 g) and 1 cabbage (1248 g), however, in the hierarchy, broccoli are classified as cabbage.

---

[5] cabbage is not a subclass of inflorescentvegetable (drumhead cabbages are not flowers), neither is inflorescentvegetable a subclass of cabbage (artichokes are inflorescent vegetables).

We get the following inconsistent conjunction of constraints:

$$(\text{cabbage}_g = \text{broccoli}_g)$$
$$\wedge (\text{cabbage}_g = 1248)$$
$$\wedge (\text{broccoli}_g = 91)$$

This problem is solved by introducing an additional variable $\text{food1}'_g$ that takes the value given in the recipe for food1 and that is handled as a specific food:

$$\text{food1}_g = \text{food1}'_g + \text{food2}_g$$

if food2 is the only food generalised by food1.

The term $\text{food1}'_g$ is also included in the equalities that give the values of food_g for any food that generalises food1.

So, for the recipe "*Vegetables in crock pot*", the previous equalities are replaced by the following ones:

$$\text{cabbage}_g = \text{cabbage}'_g + \text{broccoli}_g$$
$$\text{cabbage}'_g = 1248$$
$$\text{broccoli}_g = 91$$

which are consistent ($\text{cabbage}_g = 1248 + 91 = 1339$).

Nutritional values. We have nutritional data for each food. For example, some sugars are contained in the ingredients of the recipe, approximately 0.1 g per gram of apple, 0.12 g per gram of banana, 1 g per gram of granulated sugar, 0.05 g per gram of milk, and 0 g per gram of flour and egg. These quantities add up to give the total amount of sugars in the recipe:

$$\text{sugar}_g = 0 \cdot \text{flour}_g + 0.05 \cdot \text{milk}_g + 0 \cdot \text{egg}_g$$
$$+ 1 \cdot \text{granulatedsugar}_g + 0.1 \cdot \text{apple}_g + 0.12 \cdot \text{banana}_g$$

Formally, for a nutritional value represented by the variable nutVal, let $\alpha_{\text{food}}^{\text{nutVal}}$ be the value of nutVal contained in 1 g of food, the following equality is added to $clc(\text{DK})$:

$$\text{nutVal} = \alpha_{\text{food1}}^{\text{nutVal}} \cdot \text{food1}_g + \cdots + \alpha_{\text{foodk}}^{\text{nutVal}} \cdot \text{foodk}_g$$

Like for the constraints that encode the food hierarchy, only the most specific foods are involved in these calculations ($\text{food1}, \ldots, \text{foodk}$ are the specific ingredients of the recipe).

Representation of queries. Let $clc(\text{Q})$ be the representation of the query by a conjunction of linear constraints. The query is represented in Q by "Boolean" constraints, but we do not know how to express adequately all these constraints as linear constraints in the formalism with numerical variables used here.

A term $\neg$food in Q which stands for the request by the user not to include the food food in the recipe, can be expressed by the constraint $\text{food}_g = 0$.

By contrast, it is less easy to translate a positive literal food of the query Q. Indeed, the constraint $\text{food}_g > 0$ is not allowed in the language, since strict constraints (involving non closed subsets of $\mathcal{U}$) are not allowed in the language.

To get interesting results, we use the ingredient substitutions obtained from the basic adaptation step. For each substitution $a \rightsquigarrow b$, the following formula is added to $clc(\text{DK})$:

$$s = a_g + b_g$$

where s is a new variable. And the following formula is added to $clc(Q)$:

$$a_g = 0$$

Assume that we give more importance to the change in the value of s than to the change in the value of a_g and b_g. Thus, with no other constraint interfering, the minimal change between a value $u \in \mathcal{U}$ that satisfies $a_g = x$ ($x \in \mathbb{R}$) and $b_g = 0$, and the set of values $v \in \mathcal{U}$ that satisfy $a_g = 0$ is reached for values satisfying $b_g = x$. The expected result of the adaptation results is a "compensating" effect of the adaptation.

Choice of a distance. As mentioned in paragraph "Representation formalism" the recipes are represented in the space $\mathcal{U} = \mathcal{U}_1 \times \cdots \times \mathcal{U}_n$ where each component $\mathcal{U}_i$ is the range of the variable x_i. We consider a distance defined for any $x, y \in \mathcal{U}$ with $x = (x_1, \ldots, x_n)$ and $y = (y_1, \ldots, y_n)$ by:

$$d(x, y) = \sum_{i=1}^{n} w_i |y_i - x_i|$$

where the coefficients w_i are to be chosen.

Some criteria guide the choice of the coefficients w_i:

- Only quantities of ingredients in grams are taken into account, so for any i such that x_i gives the amount of an ingredient in some other unit than grams, $w_i = 0$.
- The coefficient for general foods must be larger than the sums of the specific foods it generalises. Given a food GenFood that generalises the foods $\text{food}1, \ldots, \text{food}k$, let $x_{i0} = \text{GenFood}_g$ be that amount in grams of GenFood and $x_{i1}, \ldots, x_{ik}$ the amounts in grams of respectively $\text{food}1, \ldots, \text{food}k$, then:

$$w_{i0} \geq w_{i1} + \cdots + w_{ik}$$

This condition ensures the "compensation" effect of the adaptation, it makes the substitution of the specific food $\texttt{food}j$ by another $\texttt{food}j'$ less important than the reduction of the amount of $\texttt{GenFood}$.

In practice, the coefficients of d are calculated thanks to the recipe base, w_i equals to the number of recipes using the food corresponding to x_i: $w_i = \mathcal{N}(x_i)$, with the notation used in Sect. 2.1.

Computation. The calculus of the adaptation of $clc(\text{R})$ to solve $clc(\text{Q})$ can be seen as an optimisation problem:

$$\text{find } y \text{ such that } x \text{ and } y \text{ satisfy } clc(\text{DK})$$
$$x \text{ satisfies } clc(\text{R})$$
$$y \text{ satisfies } clc(\text{Q})$$
$$d(x, y) \text{ is minimal}$$

The constraints that define $clc(\text{DK})$, $clc(\text{R})$ and $clc(\text{Q})$ are linear constraints but the function to be minimised, d, is not linear, so the minimisation of d under these constraints is not a linear optimisation problem. However an equivalent linear optimisation problem can be built. This new problem is defined in the space $\mathcal{U}^3 = \mathcal{U} \times \mathcal{U} \times \mathcal{U}$. Intuitively, for any $(x, y, z) \in \mathcal{U}^3$, x will be an element from $clc(\text{R})$, y an element from $clc(\text{Q})$ and $z = (z_1, \ldots, z_n)$ will be the difference per variable between values of x and y.

In addition to the linear constraint from $clc(\text{R})$ applied on x, from $clc(\text{Q})$ applied on y and $clc(\text{DK})$ applied on both x and y, the following constraints are added for every $1 \le k \le n$

$$z_k \ge y_k - x_k \quad \text{and} \quad z_k \ge x_k - y_k \quad \text{(i.e. } z_k \le |y_k - x_k|)$$

The function to minimise is $f : (x, y, z) \mapsto \sum_{i=1}^n w_i z_i$, f is a linear function over $\mathcal{U}^3$ to $\mathbb{R}$, so the minimisation of f under the constraints stated above is a linear optimisation problem. It can be noted that as no other constraints are applied to the z_k than the ones stating that $z_k \le |y_k - x_k|$, minimal values of f are reached for (x, y, z) such that $z_k = |y_k - x_k|$ and so $f(x, y, z) = d(x, y)$ when (x, y, z) is an optimum. This shows we have indeed an equivalent linear problem.

This reduction to a linear optimisation problem made it possible to use common linear optimisation tools to perform the quantity adaptation. As variables can either take real or integer values, the resulting problem is a mixed linear optimisation problem, this class of problem is NP-hard (P if all the values are reals) but the problem sizes encountered for recipe adaptation are small and manageable. In practice we use LP_Solve.[6]

---

[6] http://lpsolve.sourceforge.net/5.5/

2.2.4 Textual Adaptation of Preparations

Not all ingredients are cooked in the same way. Therefore, when an ingredient a is replaced with an ingredient b in a recipe, it may prove necessary to further adapt the preparation part. The preparation text is adapted by replacing the sequence of actions applied to a with a similar sequence of actions that is found in a different recipe where it is applied to b. Using an existing recipe guarantees that the actions used are a correct way of preparing ingredient b. Each different sequence of actions that is found to be applied to b in a recipe is called a "prototype" of b—that is, it defines one of the potentially many ways in which b can be prepared.

Consider the case of a user making the request shown in Fig. 2. No dessert recipe with fig and rice exists in the recipe base used by TAAABLE, but there is a recipe called "Glutinous rice with mangoes". The mango prototype used in this recipe is characterised by the set of actions {chill, peel, slice, remove-pit, place}. TAAABLE will suggest replacing mangoes with figs, which seems to be a satisfactory adaptation, except the need for peeling and pitting figs, that makes no sense.

In the recipe base, there are two recipes with figs. Recipe #1163 sports a {halve, sprinkle-over, dot-with, cook, brown, place} fig prototype, and #53 a more modest {cut, combine}. In order to simplify the processing and avoid discriminating near-synonyms, classes of similar actions can be grouped using an action hierarchy. This could have the effect of grouping, e.g. peel with remove-pit, cut with slice, and cook with brown.

To select the more appropriate fig prototype, the search space is organised as a concept lattice built using formal concept analysis [13] from the fig prototypes. The mango prototype is then merged into it as well. This approach is similar to techniques used in document retrieval by Carpineto and Romano [4], where a lattice is built according to keywords found in documents and a query is merged in it (see also [20, 21]). The lattice is built from a formal binary context, which maps the set of fig prototypes from recipes #53 and #1163 (the "objects") to the set of culinary actions (the "attributes"), indicating whether a given action occurs in a given prototype. The query, consisting in an additional mango prototype object from the retrieved recipe, is merged into this binary context. The resulting formal binary context is shown in Table 1, and the corresponding lattice in Fig. 7.

All prototypes having actions in common will appear together in the extent of at least one concept. The "lower" this concept is in the lattice, the more attributes the prototypes have in common. A set of candidate fig prototypes is taken from the

Table 1 Formal binary context of figs and a mango

	cut	halve	remove	place	pour	dot	cook	chill	combine
fig_#1163		×		×	×	×	×		
fig_#53	×								×
mango	×		×	×				×	

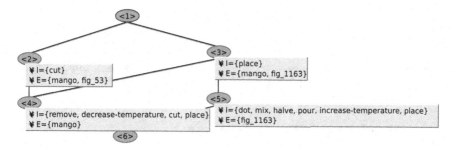

Fig. 7 Concept lattice corresponding to the binary context of Table 1. Each node corresponds to a formal concept which is composed of an extent, i.e. the objects which are instances of the concept, and an intent, i.e. the attributes composing the description of the concept. The extent is the maximal set of objects sharing the attributes in the intent, and reciprocally

extent of all concepts "immediately above" the lowest concept with mango in its extent (actually called the object concept of mango [13]). Whenever there is more than one candidate, the one that minimises the distance between its attributes and the mango's attributes is selected. In the example, fig_#53 is selected since, while both fig_#53 and fig_#1163 appear directly above mango, replacing mango with fig_#1163 would require removing three actions and adding four, whereas replacing it with fig_#53 only requires removing three actions.

The process is then completed by replacing the textual parts of the retrieved recipe dealing with mangoes with the parts of recipe #53 dealing with figs:

> [...] Blend the sauce ingredients in a pot and heat until it just reaches the boiling point. Let cool. ~~Peel the mangoes, slice lengthwise and remove the pits.~~ Cut figs into wedges. Divide the rice mixture among 6 plates. [...]

3 Managing the TAAABLE Knowledge Containers

3.1 An Ontology of the Cooking Domain

The "cooking ontology" $\mathcal{O}$ defines the main classes and relations relevant to cooking. $\mathcal{O}$ is composed of six hierarchies:

- a *food* hierarchy, related to ingredients used in recipes, e.g. Vegetable, Fruit, Berry, Meat, etc. (a part of this hierarchy is given in Fig. 8)
- a *dish type* hierarchy, related to the types of dish, e.g. PieDish, Salad, Soup, BakedGood, etc.
- a *dish moment* hierarchy, related to the time for eating a dish, e.g. Snack, Starter, Dessert, etc.
- a *location* hierarchy, related to the origins of recipes, e.g. FrenchLocation, AsianLocation, MediterraneanLocation, etc.

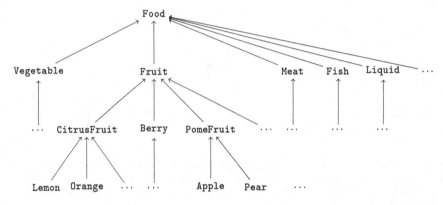

Fig. 8 A part of the food hierarchy

- a *diet* hierarchy, related to food allowed or not for a specific diet, e.g. Vegeta-rian, NutFree, etc.
- and an *action* hierarchy, related to cooking actions used for preparing ingredients, Cut, Peel, etc.

Given two classes B and A of this ontology, A is more specific than B, denoted by "AB", iff the set of instances of A is included in the set of instances of B. For instance, "CitrusFruit (referring to citrus fruits) is more specific than Fruit (referring to fruit)" means that every citrus fruit is a fruit.

3.1.1 Ontology Engineering

The ontology $\mathcal{O}$ was built in order to help the design of the retrieval and adaptation processes of the TAAABLE system. Therefore, the conceptual choice for the ontology development was strongly driven by the goal of this particular CBR system. The reuse of existing ontologies was carefully examined but no more considered as they did not cover what was intended to be reached in this project. So, after identification of the main classes during the elaboration of the cooking conceptual model, a fine-grained structuration of these classes has been carried out according to existing terminological resources and manual expertise, as it is done in object-oriented programming when the hierarchy of classes is designed.

Food hierarchy. The first version of Food hierarchy was built manually starting from several web resources such as the Cook's Thesaurus,[7] a cooking encyclopedia that covers thousands of ingredients, and Wikipedia. The main task was to select the set of relevant classes and to organise them according to the relation. At the same time, a first version of the terminological database was built in order to associate to each class (e.g., BokChoy) a linguistically preferred form (e.g., *bok choy*) as well

---

[7] http://www.foodsubs.com

as a set of lexical variants which can be a morphological variants or synonyms (e.g., *pak choy, Chinese cabbage, Chinese mustard cabbage*, etc.).

The first version of the food hierarchy and of the terminological database have then been enriched by adding iteratively new classes that occur in the recipe book but were missing in the hierarchy. For this, a semi-automatic process has been designed. This process retrieves the ingredient lines of the recipe book that cannot be linked to food classes by the annotation process. A manual expertise is required in order to determine the reason of these failures and to correct what is required. Three cases have been identified:

- the food class is missing in the hierarchy: the expert has to create it and to attach it to the most specific class(es) subsuming it.
- the food class exists in the hierarchy but the lexical form that appears in the recipe is not in the terminological database: the expert has to add this new lexical form as a lexical variant attached to the class.
- the ingredient line contains an error (e.g. no food is mentioned, a misspelling error occurs, etc.): the expert has to correct the ingredient line.

This approach has been integrated in the wiki: when a new recipe is added, the annotation process directly gives feedback to the user, so that she can correct and complete the knowledge.

Finally, some information and knowledge about ingredients is automatically collected using Freebase,[8] a RDF database. This database is queried for collecting a short description, a link to the wikipedia page, some lexical variants in several languages, the compatibility/incompatibility with some diets, and possible images.

The food hierarchy is the largest hierarchy. Its current version contains about 3000 classes, organised around nine levels. The terminological base related to food contains about 4200 English lexical forms, without taking into account singular/plurial variations.

Dish type and dish origin hierarchies. Starting from the organisation of dish types and dish origins in the Recipe Source database,[9] a list of general dish types and dish origins has been collected, and hierarchically organised following the subsumption relation.

The dish origin hierarchy contains 41 classes organised around two levels. The first level classifies the origin following their regions, such as AfricanLocation, AsianLocation, EuropeanLocation, etc. Each first level class is specialised, at the second level, by the country, origin of the dishes. For example, FrenchLocation, GermanLocation, SpanishLocation, etc. are subclasses of EuropeanLocation.

The dish type hierarchy, containing 69 classes, is organised around three levels. At the first level, there are classes like BakedGood, Burger, Dessert, MainDish, etc. The second and third levels introduce, if necessary, more specialized classes. For example, BakedGood is specialised into Bagel, Biscuit, Bread, Muffin,

---

[8] http://www.freebase.com

[9] http://www.recipesource.com

Cookie, etc. However, these classes are no more deeply detailed even if more specific categories exist in Recipe Source.

Dish moment and diet hierarchies. These two hierarchies have only one level, each concept being directly under the root (i.e. the most general concept) of the hierarchy. There are currently eight dish roles (e.g. Snack, Starter, etc.) and seven types of diet (e.g. Vegetarian, NutFree, etc.) in $\mathcal{O}$.

Action hierarchy. The Action hierarchy was obtained by organising a list of verbs automatically extracted in recipe preparation texts. The action hierarchy contains 449 classes organised around five levels. Each class is described by syntactic and semantic properties, which make the automatic case acquisition process described hereafter possible. For example, ToCut is a subclass of ToSplit and is described by the linguistic property isADirectTransitiveVerb, meaning that a direct object has to be searched in the text by the preparation annotation process. ToSplit is also described by the functional property isAnUnionVerb, meaning that this action produces more than one output.

3.1.2 Nutritional Data and Weight Equivalence Acquisition

Using the USDA Nutrient database,[10] food classes are linked to their nutritional values (sugar, fat, protein, vitamins, etc.) and some weight conversion equivalences, two types of knowledge required for the adaptation of ingredient quantities (cf. Sect. 2.2.3). For linking nutritional values and weight conversions to their corresponding ontology concepts, a semi-automatic alignment was processed. This alignment makes the link between the food label in the USDA Nutrient database and its label in the wiki as a category. As the mapping is incomplete, some of the ingredients in the wiki do not possess such information.

Figure 6 shows an example of nutritional data related to apples and bananas.

3.2 Adaptation Knowledge

TAAABLE uses a particular form of adaptation knowledge (AK): adaptation rules about the substitution of some ingredients by others (e.g. in *"My Strawberry Pie"* recipe, Strawberry could be replaced with Raspberry). Formally, an adaptation knowledge unit is a 4-tuple (*context, replace, with, provenance*), where:

- *context* represents the recipe or the class of recipes on which the substitution can be applied. An AK unit is specific if its *context* is a single recipe and generic if its *context* is a class of recipes (a specific type of dish, for example).
- *replace* and *with* are respectively the set of ingredients that must be replaced and the set of replacing ingredients.

---

[10] http://www.nal.usda.gov

- *provenance* is the source the AK unit comes from. Currently, four sources are identified:

1. TAAABLE, when AK results from a proposition of adaptation given by the reasoning process of TAAABLE.
2. *AK extractor* (resp. *Generic AK extractor*), when AK results from the specific (resp. generic) knowledge discovery system integrated in WIKITAAABLE (see Sect. 3.5.2).
3. *user*, when AK is given by a user editing the wiki, as it is usually done in cooking web site, when users add comments about ingredient substitution in a recipe. See, for example, http://en.wikibooks.org/wiki/Cookbook:substitutions.
4. *recipe*, when the AK is directly given by the original recipe when a choice between ingredients is mentioned (e.g. "100 g butter or margarine"). This particular substitutions are taken into account by a wiki bot which runs through the wiki for automatically extracting them.

According to this definition, (*"My Strawberry Pie"*, *Strawberry*, *Raspberry*, TAAABLE), is an AK unit obtained from TAAABLE, meaning that strawberries can be replaced with raspberries in the *"My Strawberry Pie"* recipe. In WIKITAAABLE, each substitution is encoded as a wiki page like the one given in Fig. 11. A semantic query is used to feed automatically the *Substitutions* section of a recipe page, as visible in Fig. 9.

3.3 WIKITAAABLE: *A Semantic Wiki for* TAAABLE

WIKITAAABLE[11] is a semantic wiki which is used in order to represent, edit and maintain knowledge used by TAAABLE [7]. To develop WIKITAAABLE, we relied on an existing tool: Semantic MediaWiki [17].[12] Semantic MediaWiki is an extension of MediaWiki (a well-known wiki engine, used among others by Wikipedia) enabling users to embed semantic in their wiki pages. We decided to use a wiki to benefit from the online and collaborative edition facilities it provides. Semantic MediaWiki was then a suitable solution to enable the introduction of knowledge units in the wiki.

Using WIKITAAABLE, users can browse, query, and edit the knowledge base through a user-friendly interface. The CBR engine is connected to WIKITAAABLE and uses the knowledge base during reasoning process. WIKITAAABLE embeds bots (small programs) which are in charge of performing automated tasks such as annotation of pages, tests or maintenance operations. Finally, additional interfaces are implemented within WIKITAAABLE so that users have only one tool to master when they manage knowledge. In summary, WIKITAAABLE acts as a collaborative and shared space between humans and agents.

WIKITAAABLE is composed of four main components:

---

[11] http://wikitaaable.loria.fr/

[12] http://semantic-mediawiki.org

Fig. 9 An example of a WIKITAAABLE recipe: "My strawberry pie"

Semantic MediaWiki.   Semantic MediaWiki is the base of the architecture of WI-
    KITAAABLE. Users access the system through the web interface. The CBR engine
    and bots are connected to the wiki through a set of predefined semantic queries.
    Semantic MediaWiki encodes the knowledge base as a set of semantic wiki
    pages. The indexed recipes, the ontology classes, and the adaptation knowledge
    are encoded in semantic wiki pages. See Figs. 9–11 for examples of wiki pages
    encoding respectively a recipe, an ontology class, and an adaptation.
Import bots.   Scripts import TAAABLE knowledge base, and especially the recipes
    provided by the CCC into Semantic Media Wiki. These bots are only run once to
    bootstrap the wiki by creating a minimal set of knowledge units and recipes.
Recipe Annotation Bot.   The recipe annotation bot parses the recipe pages, extracts
    ingredient information, and updates recipe pages with semantic annotation and
    categorisation of recipes. The parsing and update of recipes is done using the
    mediawiki API, accessing the knowledge base is done using predefined semantic
    queries. This bot is triggered each time a new recipe is added into the wiki in
    order to build its semantic annotation, and each time a new ingredient is added
    into the wiki in order to fill the missing annotations in.

Fig. 10 An example of an
ontology class: the `Berry`
food class

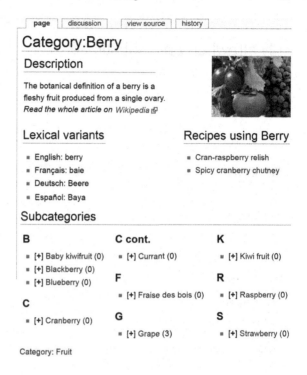

| page | discussion | view source | history |

Category:Berry

Description

The botanical definition of a berry is a
fleshy fruit produced from a single ovary.
Read the whole article on Wikipedia

Lexical variants

- English: berry
- Français: baie
- Deutsch: Beere
- Español: Baya

Recipes using Berry

- Cran-raspberry relish
- Spicy cranberry chutney

Subcategories

B
- [+] Baby kiwifruit (0)
- [+] Blackberry (0)
- [+] Blueberry (0)

C
- [+] Cranberry (0)

C cont.
- [+] Currant (0)

F
- [+] Fraise des bois (0)

G
- [+] Grape (3)

K
- [+] Kiwi fruit (0)

R
- [+] Raspberry (0)

S
- [+] Strawberry (0)

Category: Fruit

Knowledge Acquisition Interfaces.   WIKITAAABLE implements several knowledge
acquisition interfaces. See Sect. 3.5.2 for illustration and details.

3.4 Case Acquisition from Texts

3.4.1 Ingredient Annotation

The case base engine requires a formal representation of a recipe. The annotation
process aims at formally encoding using semantic wiki properties and classes the
content of a recipe as well as meta-properties like, for example, its origin and the type
of dish the recipe will produce. This process is in between *controlled indexing* [16]
where terms come from a predefined terminology and *semantic annotation* [27]
where terms (named entities, sequences of words) are explicitly associated with the
respective and most specific classes in the ontology. The result of the annotation of
the ingredient part of a recipe is a set of food classes linked each to a recipe with
the `hasIngredient` semantic wiki property. For that, the list of ingredients is
parsed. First, each ingredient entry in the recipe is parsed and split into the following
4-tuple (`<quantity>,<unit>,<foodcomponent>,<modifiers>`). For

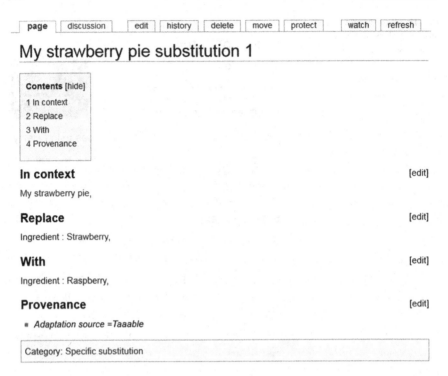

Fig. 11 An example of an adaptation rule page where `Strawberry` is replaced with `Raspberry` in the "My Strawberry Pie" recipe

example, the entry "`1/3 cup milk`" is parsed as (`1/3,cup,milk,_`). The terminological database guides the parsing process: the lexical variants are used conjointly to regular expressions for searching in the text an instance of a class of $\mathcal{O}$. For example, as "*pak choy*" is a lexical form associated to the food class `BokChoy` in the food hierarchy, the entry "*1 kg sliced pak choi*" is parsed as (`1, kg, BokChoi, sliced`) and the recipe containing this ingredient line is linked to the `BokChoy` food class using the `hasIngredient` property.

For example, the recipe entitled "*My strawberry pie*", illustrated in Fig. 9, is indexed by the conjunction of the ingredients `Water`, `Sugar`, `PiePastry`, `Strawberry`, `Cornstarch`, and `CoolWhip`.

3.4.2 Annotation of the Dish Types and Dish Origins

Another annotation process indexes recipes w.r.t. its origin (e.g., `AsianLocation`) and w.r.t. the type(s) of dish produced (e.g., `MainDish`, `Dessert`). As there is no indication about this in the recipe book, Recipe Source is used again in order to build a corpus where recipes are assigned to their origin and dish type(s). The process to determine the origin and the dish type of a recipe is based on 3 steps. For a given recipe:

- If there exists a recipe in Recipe Source with the same title, then the origin and the dish type(s) of the recipe in Recipe Source are assigned to the Recipe Book recipe;
- If the title of the recipe (e.g., *"Chinese Vegetable Soup"*) contains keywords corresponding to subclasses of DishType (e.g., soup) or DishOrigin (e.g., *"Chinese"*) then these origin and dish type(s) are assigned to the recipe;
- A set of association rules has also been extracted from the Recipe Source corpora, using the data-mining toolkit CORON [26]. According to exact associations rules (with 100 % confidence) of the form <set of ingredients> $\longrightarrow$ <origin or dish type> (e.g., vanilla bean, banana, chocolate $\longrightarrow$ dessert), assignations of origin and dish type(s) can be done as follows: if part of the recipe matches the left-hand side of the rule, then the origin and/or the dish type(s) in the right-hand side is assigned to the recipe.

For example, the recipe entitled "My strawberry pie" is indexed by some ingredients, as presented below, but also by the following dish types: Dessert and PieDish.

As the semantic representation of a recipe is stored into a wiki, users are involved for correcting manually the annotation. Users may enter new ingredients, new lexical variants or also choose the dish types, the dish origins, the dish moments associated to recipes.

3.4.3 Preparation Annotation

To make possible the textual adaptation task described in Sect. 2.2.4, a more complete formal case representation, including the way the ingredients are prepared, is required. Because it consists, in a nutshell, in combining a set of ingredients in specific ways until an edible dish is obtained, it is natural to represent a recipe as a tree, as shown in Fig. 12. Each node represents the state of a food component (an ingredient or a mix of ingredients) at a given time, and each edge $\langle a, b \rangle$ represents an action applied on a food component a that yields a new food component b. A set of edges sharing the same head (b) represents a "combining action", i.e. an action applied to many food components at once that yield out one new food component.

This representation is built iteratively, verb by verb, following the order of the text. Some sentences are difficult to analyse, and the partially built representation can be helpful. For example, in a sentence such as "Peel the mangoes, slice [the mangoes] lengthwise and remove the pits [from the mangoes]", it is easy for humans to understand that mangoes are being sliced and pitted, but some heuristics are needed for making this understandable to a computer. Each action is assigned an arity, making it possible to detect the absence of an argument. Whenever this happens, it is assumed that the missing argument corresponds to the last node that was added to the tree.

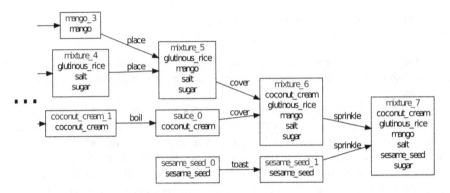

Fig. 12 Excerpt of the formal representation of "Glutinous rice with mangoes"

This is one way of dealing with anaphoras, i.e. the phenomenon wherein a different word—or no word at all as it might be—is used to represent an object.

Other types of anaphora appear as well. Still in "Glutinous rice with mangoes", the expression "seasonings ingredients" is found, referring to some set of food components. The ingredient hierarchy is used to find all the nodes of the tree that fit under the "seasonings" category. A phrase such as "cover with sauce" is trickier because there is no obvious clue in the text or in the ontology about which food component this "sauce" may be. We built, from the analysis of thousands of recipes, "target sets" of ingredients that usually appear in the food components being referred to by word such as "sauce" or, say, "batter". Because a quantity of recipes include coconut cream-based sauces (and none contain, say, rice-based sauces), it makes sense to assume that, if a food component in a given recipe contains coconut cream, this is the one indicated by the use of the word "sauce".

Once the tree representation is built, it is straightforward to identify the sequence of actions being applied to a given ingredient in a recipe, in order to replace it as described in Sect. 2.2.4. Because we defined the prototype of an ingredient as the sequence of actions applied to this ingredient, up to and including the first combining action, in practice, only the leftmost part of the tree is replaced or used as a replacement. This may mean missing relevant actions, but also getting rid of actions that, while being applicable to certain mixtures containing an ingredient, may not make sense when applied directly to this ingredient. In the mango rice, for example, the *chill, peel, slice lengthwise, remove pits*, and *place on top* actions are kept, whereas *cover with, sprinkle with*, and *serve* are lost. It would seem that those last three actions are indeed too generic to be relevant in the present application. As an added benefit, because language processing errors accumulate, the leftmost parts of the tree, corresponding to the beginning of the recipe, are built with a higher reliability than the rightmost parts at this time.

3.5 Adaptation Knowledge Acquisition (AKA)

This section presents the various AKAstrategies implemented in TAAABLE. When TAAABLEadapts a recipe to match a user query, it produces an ingredient substitution. If users are satisfied with the adapted recipe, e.g. if they click on the "OK" button to validate the adaptation, the ingredient substitution is stored in WIKITAAABLEas an AKunit for future reuse. This AKacquisition strategy is straightforward but proves to be efficient.

If users are not satisfied with the adapted recipe, they can click on the "NOT OK" button. A process aiming at repairing the failed adaptation is then triggered. If this process is successful, additional AKunits are acquired. This is what we call failure-driven AKA. This process is described hereafter.

We have also implemented other mechanisms to support the process of acquiring AKindependently of a specific query. For that, we have tuned well-known knowledge discovery (KD) techniques. These mechanisms are described in the second part of this section.

3.5.1 Failure-Driven Adaptation Knowledge Acquisition

The TAAABLE system has been extended in 2009 to support online AKA [3]. The knowledge acquisition process complies with the FIKA principles of knowledge acquisition in case-based reasoning [6], in which the knowledge is acquired:

- **online**: in a particular problem-solving context,
- **interactive**: knowledge is acquired through interaction with users, and
- **opportunistic**: knowledge acquisition is triggered in response to a reasoning failure, i.e. when the user is not satisfied with the proposed solution.

Two strategies have been implemented to repair an adaptation failure, which differ by the source of AK. In the first strategy, AKis acquired only from the user and from the domain knowledge. In the second strategy, a KDstep is added to the knowledge acquisition process. When the KDprocess, called CABAMAKA [2, 8], is triggered, it turns the case base into an additional source of AK.

The different steps involved in the overall AKAprocess are summarised in Fig. 13. The knowledge acquisition process starts at the solution test phase of the CBR cycle, when the user is not satisfied with the proposed solution. Repairing a failed adaptation is a two-step process. First, an explanation of the failure is identified through interactions with the user and a predefined explanation pattern is selected and instantiated. Then, the user is asked to choose among as set of repair strategies that are associated to this failure explanation pattern. This approach is inspired from CHEF's critic-based adaptation [14]. CHEF's approach to adaptation is to run a simulation and use a causal model to detect and explain potential problems in the generated solution. To solve these problems, CHEF makes use of a set of critics. A critic identifies a set of potential problems that can occur in the adapted solution and associates to them a repair strategy.

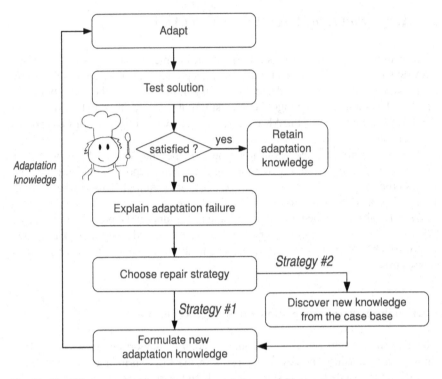

Fig. 13 The different steps of the interactive knowledge acquisition process

The knowledge discovery process is optionally triggered when a repair strategy has been chosen. Its goal is to search the case base for information needed to instantiate this repair strategy, thereby implementing a case-based substitution approach in the spirit of DIAL's *adaptation=transformation+memory search* approach [18]. In the DIAL system, rule-based adaptation is performed by selecting an abstract rule and by searching for information needed to instantiate the rule. In our approach, a repair strategy defines a set of constraints on the adaptation rule to apply. These constraints are used to restrict both the training set and the hypothesis space of the learning process. CABAMAKA extracts from the case base a set of adaptation rules that satisfy these constraints. AK is dynamically generated from the case base and transferred to the adaptation knowledge base. The KD process is designed to only generate adaptation rules that are useful for solving a particular adaptation problem. Since the extracted adaptation rules are in a format directly usable by the system, the system can use them to re-run the adaptation step and propose a new solution to the user. The validation step is performed online by the system user, which is presented with the repaired solution together with the AK that was used to generate the solution.

Once the user is satisfied with the proposed solution, the discovered AK is retained in the AK base for future reuse.

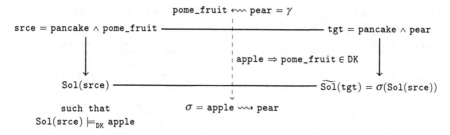

Fig. 14 An adaptation proposed by the TAAABLE system to answer the query "I want a pear pancake". A solution $\widetilde{Sol}(tgt)$ for the target problem tgt is constructed from the representation Sol(srce) of the retrieved recipe *Apple pancakes from the townships*. The substitution $\sigma =$ apple $\rightsquigarrow$ pear is automatically generated from the substitution by generalisation $\gamma =$ apple $\rightsquigarrow$ pome_fruit using the axiom apple $\Rightarrow$ pome_fruit of DK

In the remaining of this section, the two AKA strategies are illustrated. In the solution test phase, the retrieved recipe Sol(srce) is presented to the user together with the adaptation path AP that was used to generate the candidate solution which is denoted $\widetilde{Sol}(tgt)$. In this example, the recipe *Apple Pancakes from the Townships* is presented to the user in the solution test phase, together with the adaptation path AP $= \sigma$ (with $\sigma =$ apple $\rightsquigarrow$ pear), that was used to generate the solution $\widetilde{Sol}(tgt)$ (Fig. 14).

In the failure explanation step, the user is encouraged to formulate an explanation of the adaptation failure. The failure explanation step is achieved in three substeps:

- *Substitution Selection.* The user selects a substitution $\sigma = A \rightsquigarrow B$ of AP which is problematic, where A and B are conjunctions of literals. In the example, the user selects the substitution $\sigma =$ apple $\rightsquigarrow$ pear, so A $=$ apple and B $=$ pear.
- *Explanation pattern selection.* The user selects an explanation pattern. Each explanation pattern explains the failure of a single substitution σ of the adaptation step, in which the solution Sol(pb) of a problem pb is transformed in a solution $\sigma(Sol(pb))$. So far, three explanation patterns have been considered in the TAAABLE system: (1) an ingredient x of B requires an ingredient y which is not in $\sigma(Sol(pb))$, (2) an ingredient x of Sol(pb) requires an ingredient y of A which has just been removed, and (3) an ingredient x of B is not compatible with an ingredient y of $\sigma(Sol(pb))$. Each one expresses a dependence between ingredients that was violated in the proposed recipe $\sigma(Sol(pb))$. In this example, the explanation pattern selected by the user expresses the fact that an ingredient x was added to the recipe after applying the substitution σ is incompatible with one of the ingredients of the recipe $\sigma(Sol(pb))$.
- *Explanation Pattern Instantiation.* The user selects the ingredients x and y in a list of propositions. In the example, the user selects the ingredients $x =$ pear and $y =$ cinnamon, in order to express the fact that keeping the cinnamon in the retrieved recipe is not satisfactory when pears are added.

Table 2 An example of failure explanation pattern and the associated repair strategies that are used to repair an ingredient substitution rule $\sigma = A \rightsquigarrow B$ in the cooking domain

Failure explanation	Associated repair strategies
An ingredient x of B is incompatible with an ingredient y of the adapted recipe	–if $\mathtt{tgt} \not\models_{DK} x$, remove x –if $\mathtt{tgt} \not\models_{DK} x$, find a substitute for x –if $\mathtt{tgt} \not\models_{DK} y$, remove y –if $\mathtt{tgt} \not\models_{DK} y$, find a substitute for y
$x \leftarrow \mathtt{pear}$ $y \leftarrow \mathtt{cinnamon}$ ($\mathtt{pear}$ is incompatible with $\mathtt{cinnamon}$ in the recipe "*Apple pancakes from the townships*")	–remove $\mathtt{cinnamon}$ –find a substitute for $\mathtt{cinnamon}$

To each explanation pattern is associated a set of repair strategies. A repair strategy is selected by the user among the ones that are applicable.[13] In the example, four repair strategies correspond to the selected explanation pattern but only the last two ones are applicable since $\mathtt{tgt} \models_{DK} \mathtt{pear}$ (Table 2). The two repair strategies proposed to the user are:

Strategy#1: remove $\mathtt{cinnamon}$
Strategy#2: find a substitute for $\mathtt{cinnamon}$

Two types of repair strategies are considered: the strategies that consist in adding or removing an ingredient, and the strategies that consist in finding a substitute for an ingredient.

Strategies that consist in adding or removing an ingredient are easy to instantiate. To add (resp. remove) an ingredient, the ingredient has to be selected on a list for addition (resp. removal). A repaired substitution σ' is generated from the substitution $\sigma = A \rightsquigarrow B$, in which an ingredient has been added or removed. In the example, the substitution σ' generated to repair σ in Strategy #1 is:

$$\sigma' = \mathtt{apple} \wedge \mathtt{cinnamon} \rightsquigarrow \mathtt{pear}$$

To instantiate the strategies that consist in finding a substitute for an ingredient, the CABAMAKA KDprocess is run in an opportunistic manner. Its goal is to learn a substitution σ' from the comparison of two recipe sets. In the example, the substitution σ' is of the form:

$$\sigma' = \mathtt{apple} \wedge \mathtt{cinnamon} \wedge something_1 \rightsquigarrow \mathtt{pear} \wedge something_2$$

---

[13] It can be noticed that at least one repair strategy is always applicable for a selected explanation pattern. Indeed, if the explanation pattern corresponds to a dependence of the form "x requires y", then either $\mathtt{tgt} \models_{DK} y$ or $\mathtt{tgt} \not\models_{DK} y$ holds. If the explanation pattern corresponds to a dependence of the form "x and y are incompatible", then $\mathtt{tgt} \models_{DK} x$ and $\mathtt{tgt} \models_{DK} y$ holding simultaneously would mean that $\mathtt{tgt}$ contains two incompatible ingredients.

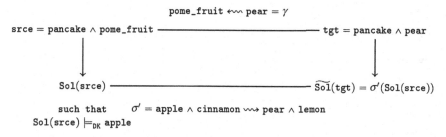

Fig. 15 A repaired adaptation proposed by the TAAABLE system to answer the query "I want a pear pancake". In this new adaptation, apples and cinnamon are substituted in the retrieved recipe by pear and lemon

in which *something*₁ and *something*₂ are conjunctions of literals that need to be determined.

The form of the σ' substitution is used to tune the KD process. In the data preparation step, σ' is used to restrict the training set. The result set is also filtered in order to retain only the adaptation rules that have the form of σ' and that are applicable to solve the adaptation problem at hand [2].

In the example, the user selects the substitution:

$$\sigma' = \texttt{apple} \land \texttt{cinnamon} \rightsquigarrow \texttt{pear} \land \texttt{lemon}$$

The substitution σ' is used to re-run the adaptation step and generate a new solution $\widetilde{\texttt{Sol}}(\texttt{tgt})$. The repaired adaptation is summarised by Fig. 15.

During the solution test phase, the retrieved recipe $\texttt{Sol}(\texttt{srce})$ is presented to the user together with the new adaptation path $\texttt{AP}' = \sigma'$ that is applied to $\texttt{Sol}(\texttt{srce})$ to produce the solution $\widetilde{\texttt{Sol}}(\texttt{tgt})$.

3.5.2 Knowledge Discovery for Adaptation Knowledge Acquisition

Two interactive and opportunistic online systems have also been implemented to acquire AK independently of any specific query. These systems are also based on KD processes. In AK EXTRACTOR and GENERIC AK EXTRACTOR systems, users query the system, and validate some extraction results, eventually with some corrections, as AK unit. Therefore, users validate knowledge on the fly, in context (i.e. for a specific recipe), which is more convenient than dealing with a huge amount of candidate knowledge (information units) out of any context (i.e. for a KD process applied to a large set of recipes).

AK EXTRACTORwas built in order to acquire specific AK units, i.e. AK units that applies only to one recipe. AK EXTRACTOR takes into account compatibility and incompatibly of ingredients, overcoming the weakness of the adaptation process of TAAABLE. AK EXTRACTOR is based on a comparison of ingredients between the

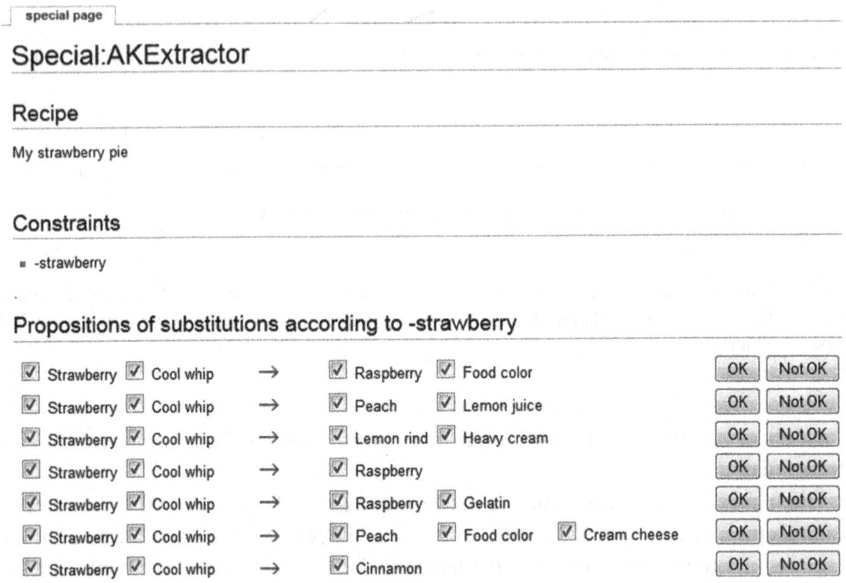

Fig. 16 Validation interface of the AK Extractor system for the query: adapt the "My Strawberry Pie" without strawberry

recipe to adapt, and a set of similar recipes. The system selects first recipes which have a minimal number of ingredients in common and a minimal number of different ingredients with the recipe that has to be adapted. Closed itemsets (CIs) [13] are extracted from variations of ingredients between recipes, starting from variations between each selected recipe of the previous step and the recipe to adapt. The CIs are then filtered and ranked with specific rules (which are detailed in [11]). The system displays the propositions for ingredient substitutions coming from the first five better ranked CIs. Figure 16 shows the validation interface of the AK Extractorsystem coming from user constraints: "adapt the recipe *"My Strawberry Pie"* without strawberry". The first proposition of substitution means that `Strawberry` and `CoolWhip` could be substituted with `Raspberry` and `FoodColor`. Finally, user may give feedback for (fully or partially) validating or not some of these substitution propositions.

Generic AK Extractor was built in order to acquire generic AK units. Generic AK units provide rules that could be applied in a larger number of situations, because the context of an AK unit will be a set of recipes (e.g. for cakes), instead being only one recipe. Generic AK Extractor defends an approach based on CIs for extracting generic substitutions (generic AK units) starting from specific ones (specific AK units) [12]. Thanks to a dedicated interface, users can trigger the extraction of CIs from the specific AK units applicable to a set of recipes. CIs extracted are then filtered and ranked with specific rules (based on two measures commonly used in KD processes based on CIs: support and stability) before being presented to the user.

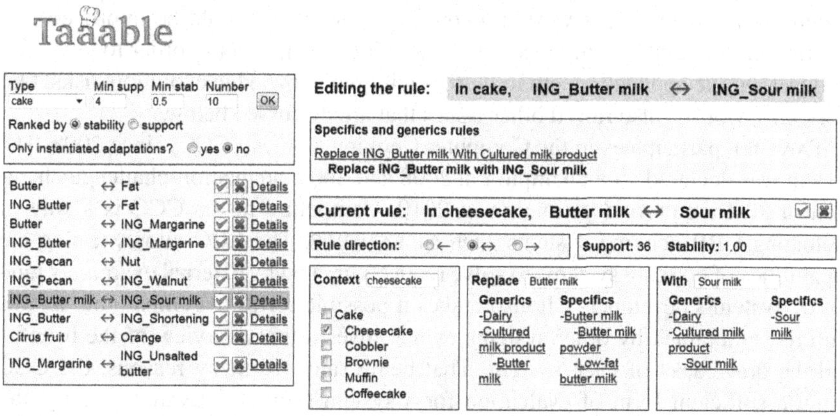

Taaable

Fig. 17 Interface of the GENERIC AK EXTRACTOR system

The system displays the propositions of ingredient substitutions. The user can validate or repair a proposition of AK in order to generalise or specialise the substituted and substituting ingredients, change the direction of the rule and/or the class of recipes for which the substitution held. For example, in Fig. 17, the system is queried in order to acquire adaptation rules for cake dishes. Ten rules are proposed by the system (ranked by stability with a minimal support of four and a minimal stability of ten). Before validating a proposition of adaptation rule, the user can change the dish type on which the rule applies, as well as ingredients involved in the rule.

For the two systems, once validated, the adaptation rule is stored as an AK unit in WIKITAAABLE in order to be used by TAAABLE for the future adaptations.

4 Conclusion and Ongoing Work

In this chapter, the TAAABLE Case-Based Reasoning system is presented. TAAABLE consists of a user interface, an *ad-hoc* CBR engine and an extensive knowledge base stored in a semantic wiki. The knowledge base consists of the recipes, the domain ontology, and a set of AK. A semantic wiki is used in order to facilitate knowledge enrichment tasks for TAAABLE users and contributors.

TAAABLE is an active research project. Improvements and new developments are made on a regular basis. Ongoing work includes the following tasks. First, TAAABLE's user interface has to be improved in order to facilitate the expression of complex queries for inexperienced users. Next, the knowledge acquisition modules have to be better integrated into the main application. This is important to encourage users to experiment with these modules. Indeed, these modules have mainly been used by TAAABLE developers, aware of how the system works; if these modules were efficiently used by more users, including, ideally, cooking experts with little skill in computer science, the TAAABLE knowledge containers would be improved. For this reason, we need to build a community of TAAABLE users in order to get better feedback on the application and to help us with the knowledge evolution tasks. Our work on TAAABLE also raised other issues that are discussed below.

TAAABLE participates in the Computer Cooking Contest (CCC) since 2008. The system was declared vice-champion in 2008 and 2009, adaptation-challenge champion in 2009 and world champion in 2010. Competing in the CCC is a way of evaluating CBR cooking systems such as TAAABLE. The CCC allows testing the capabilities of cooking systems to solve requests created by external examiners (and not by system's developers). It also makes it possible for us to compare our results with those obtained by other systems, which gives a first overview of the benefits and the drawbacks of each system. That being said, for many reasons, the CCC is not a sufficient form of evaluation for a system such as TAAABLE. Firstly, the CCC does not provide any Gold Standard or evaluation protocol. The evaluation is a comparative evaluation between systems only. Therefore, it is not possible to quantitatively assess the quality of the results obtained by a system. The CCC does not comply with what we usually expect in a benchmark. Second, the tasks given in the CCC are very general and, consequently, the evaluation is also very general. Applications are not evaluated on specific points such as quality of the knowledge acquisition, or adaptation of preparations, for example. Third, the CCC gathers only a small number of participants and experiments with systems during a very short amount of time, which does not provide good coverage of all the possible outcomes of the competing systems. We believe that elaborating evaluation strategies for the different components of TAAABLE is an important challenge. It is a complex task for two reasons. Firstly, there are a lot of components to evaluate (quality of the knowledge base, quality of the retrieval process, quality of the adaptation process, ergonomics of the user interface, etc.). Second, in the cooking domain, evaluation has a necessary degree of subjectivity. Indeed, the "quality" of a recipe produced in response to a query is very user-dependent. Therefore, we need to find a way to take into account the subjective dimension in the evaluation process. Implementing interfaces allowing users to provide feedback and the results, and being able to take into account this feedback to improve TAAABLE is a major challenge.

During the development of TAAABLE, we have extended it with many components to improve the reasoning process, the user interface, the integration of external sources of knowledge, and the interactive acquisition of knowledge. Among these improvements is WIKITAAABLE, a semantic wiki that acts as a unique tool to manage the whole knowledge base of the system. WIKITAAABLE was designed to solve the knowledge maintenance problems we had during the first year of the project. During this year, we built, mostly manually, the knowledge sources used by TAAABLE. To that end, we had to integrate external knowledge sources and to annotate the recipe book. This manual process led to many errors that had to be corrected. However, because several people were involved in this process, versioning issues soon arose. This is how we realised we needed a collaborative tool in order to manage the knowledge base of the system. We chose Semantic MediaWiki because it seemed to be a smooth solution to support collaborative editing of knowledge bases, because it was a web-based solution, and because it provided us with connectors in order to easily plug in the CBR engine of TAAABLE. But, with the ease of use, risks appeared. Indeed, with WIKITAAABLE, it is now very easy to make a "mistake" while editing the knowledge base and thus to jeopardise TAAABLE results. For that reason, we need to setup testing and validation mechanisms (especially regression tests) to ensure that ontology modifications do not degrade the performances of the system. This is an active topic of research for TAAABLE, within the framework of the Kolflow project.

Another topic we wish to investigate is the consideration of different viewpoints in a single system. For example, in Brazil, avocados are often eaten as a dessert, mixed with milk and sugar, while in France, they are eaten as a starter, with shrimps and mayonnaise. Given the way the knowledge base is organised in TAAABLE, it is currently impossible to represent these two points of view. It is always possible to define AK to solve this issue, but this solution does not provide the flexibility we expect. We believe that viewpoints must be taken into account in the design of the knowledge base, and we investigate solutions making it possible for users to define their own knowledge bases while collaborating with others users. This raises a lot of questions regarding knowledge representation, knowledge sharing, consistency of knowledge bases and conflict management, but it is a very challenging and promising topic.

Last but not least, making the CBR inference engine evolve appears to be an important issue. For the moment, the inference engine of TAAABLE performs query reformulations (generalisations) in order to find satisfactory recipes. Recipes are adapted through a specialisation process in order to match the components of the initial query. Therefore, the engine is based on a generalisation/specialisation mechanism. The engine only processes knowledge expressed in propositional logic. This representation formalism does not allow representation of attributes (such as the quantities of ingredients or their properties). As a consequence, to process such attributes, additional adaptation mechanisms have to be developed on top of the main engine. Moreover, these attributes are not taken into account during the retrieval process (for example, it is not possible to ask TAAABLE for a "low sugar recipe" at the moment). Retrieval only relies on the knowledge represented in the ingredient ontology. Additional AK, such as adaptation rules, is not taken into account during this step, but it should. We should therefore move towards a more expressive representation language while making sure that computation time remains reasonable. We need to make the inference engine evolve accordingly. Indeed, a joint and coordinated development of the knowledge bases and the reasoning engine is required. Such a refactoring would also give us the opportunity to develop a more generic inference engine and to experiment with it in other application domains.

Resumes of the Authors

Amélie Cordier   is an assistant professor at the University of Lyon 1. She does her research at the LIRIS Laboratory. She got her Ph.D. from Lyon 1 University.

Her main research field is dynamic knowledge engineering. She works with case-based reasoning and trace-based reasoning. She lead the TAAABLE project it 2008 and 2009 and she organized the Computer Cooking Contest in 2010 and 2011.

Jean Lieber   is an assistant professor of Lorraine Université with a Ph.D. and a habilitation degree in computer science, doing his research at LORIA. His main research field is CBR, with an emphasis on knowledge representation for CBR and adaptation in CBR. He has participated to the TAAABLE project since the first Computer Cooking Contest (2008) and was the TAAABLE project leader in 2011.

Emmanuel Nauer   is an assistant professor of Lorraine Université and member of the Orpailleur team, at LORIA. Emmanuel Nauer is currently the leader of the TAAABLE project, on which he has participated since its beginning. In the TAAABLE project, he has been in charge of the acquisition of the ontology, of the annotation process of recipes, and of knowledge discovery for improving the results of the CBR system.

Fadi Badra   received a Ph.D. degree in computer science from the University of Nancy in 2009. He is an assistant professor at the University of Paris 13 Bobigny, where he joined the Biomedical Informatics research group (LIM&BIO). Fadi's research contributions to TAAABLE concerned techniques to acquire adaptation knowledge, either from the end user, or from the case base by the means of knowledge discovery.

Julien Cojan   is currently an INRIA engineer in the team Wimmics. He works on data extraction from semi-structured textual resources for the semantic web. He has a Ph.D. in computer science from Nancy University in 2011, on the application of belief revision to CBR, that is used in the TAAABLE system for adapting ingredient quantities.

Valmi Dufour-Lussier   graduated from Montréal University and Nancy 2 University and is currently a Ph.D. candidate in Computer Science at Lorraine University. His area of research is located at the interface between textual CBR and spatio-temporal reasoning. He has been involved in TAAABLE since 2009, and has led the research on recipe text adaptation.

Emmanuelle Gaillard   graduated from Nancy 2 University, and is currently a Ph.D. student in Computer Science at Lorraine University. Her thesis focus on acquisition and management of meta-knowledge to improve a case-based reasoning system. She works also on adaptation knowledge discovery and applies this work to the TAAABLE system.

Laura Infante-Blanco   obtained her computing engineer degree in Universidad de Valladolid, Spain, in 2011. She is currently working as an INRIA engineer in the orpailleur team developing a generic ontology guided CBR system. She has been involved in the development of WIKITAAABLEand she is currently in charge of the wiki management.

Pascal Molli   is full professor at University of Nantes and is head of the GDD Team in LINA research center. He has published more than 80 papers in software engineering, information systems, distributed systems, and computer supported cooperative work (CSCW). He mainly works on collaborative distributed systems and focuses on algorithms for distributed collaborative systems, distributed collaborative systems, privacy and security, and collaborative distributed systems for the Semantic Web.

Amedeo Napoli   is a CNRS senior scientist and has a doctoral degree in Mathematics and an habilitation degree in computer science. He is the scientific leader of the Orpailleur research team at the LORIA laboratory in Nancy. He works in knowledge discovery, knowledge representation, reasoning, and Semantic Web.

Hala Skaf-Molli   received a Ph.D. in computer science from Nancy University in 1997. From 1998 to September 2010, she was an associate professor at Nancy University, LORIA. Since October 2010, she is an associate professor at Nantes University, LINA. She has mainly worked on distributed collaborative systems and social semantic web.

References

1. Baader, F., Hollunder, B., Nebel, B., Profitlich, H.J.: An empirical analysis of optimization techniques for terminological representation systems. In: Proceedings of the 3rd International Conference on Principles of Knowledge Representation and Reasoning (KR'92), Cambridge, Massachussetts, pp. 270–281 (1992)
2. Badra, F.: Extraction de connaissances d'adaptation en raisonnement à partir de cas. Ph.D. thesis, Université Henri Poincaré—Nancy I (2009)
3. Badra, F., Cordier, A., Lieber, J.: Opportunistic adaptation knowledge discovery. In: Proceedings of 8th International Conference on Case-Based Reasoning Research and Development, ICCBR 2009, pp. 60–74, Seattle, WA, USA, 20–23 July 2009
4. Carpineto, C., Romamo, G.: Order-theoretical ranking. J. Am. Soc. Inf. Sci. **51**(7), 587–613 (2000)
5. Cojan, J., Lieber, J.: Belief revision-based case-based reasoning. In: G. Richard (ed.) Proceedings of the ECAI-2012 Workshop SAMAI: Similarity and Analogy-Based Methods in AI, pp. 33–39 (2012)
6. Cordier, A.: Interactive knowledge acquisition in case based reasoning. Ph.D. thesis, Université Claude Bernard Lyon 1, France (2008)
7. Cordier, A., Lieber, J., Molli, P., Nauer, E., Skaf-Molli, H., Toussaint, Y.: WIKITAAABLE: a semantic wiki as a blackboard for a textual case-based reasoning system. In: SemWiki 2009—4th Semantic Wiki Workshop at the 6th European Semantic Web Conference—ESWC 2009, Heraklion, Grèce. http://hal.inria.fr/inria-00432353 (2009)
8. d'Aquin, M., Badra, F., Lafrogne, S., Lieber, J., Napoli, A., Szathmary, L.: Case base mining for adaptation knowledge acquisition. In: Proceedings of the International Conference on, Artificial Intelligence, IJCAI'07, pp. 750–756 (2007)
9. DeMiguel, J., Plaza, L., Díaz-Agudo, B.: ColibriCook: a CBR system for ontology-based recipe retrieval and adaptation. In: M. Schaaf (ed.) Workshop Proceedings of the 9th European Conference on Case-Based Reasoning, pp. 199–208, Trier (2008)
10. Developed, J.J.A., Knowledge, C.O.O.: P.J. Herrera and P. Iglesias and D. Romero and I. Rubio and B. Díaz-Agudo. In: M. Schaaf (ed.) Workshop Proceedings of the 9th European Conference on Case-Based Reasoning, pp. 209–218. Trier (2008)
11. Gaillard, E., Lieber, J., Nauer, E.: Adaptation knowledge discovery for cooking using closed itemset extraction. In: Proceedings of the 8th International Conference on Concept Lattices and their Applications—CLA 2011, pp. 87–99, Nancy, France. http://hal.inria.fr/hal-00646732 (2011)

12. Gaillard, E., Nauer, E., Lefevre, M., Cordier, A.: Extracting Generic Cooking Adaptation Knowledge for the TAAABLE Case-Based Reasoning System. In: Cooking with Computers workshop @ ECAI 2012. Montpellier, France (2012). http://hal.inria.fr/hal-00720481
13. Ganter, B., Wille, R.: Formal Concept Analysis. Springer, Heidelberg (1999)
14. Hammond, K.J.: Case-Based Planning: Viewing Planning as a Memory Task. Academic Press, San Diego (1989)
15. Hanft, A., Newo, R., Bach, K., Ihle, N., Althoff, K.D.: Cookiis—a successful recipe advisor and menu advisor. In: Montani, S., Jain, L. (eds.) Successful Case-Based Reasoning applications, pp. 187–222. Springer, Berlin (2010)
16. Jo, T.C., Seo, J.H., Hyeon, K.: Topic spotting on news articles with topic repository by controlled indexing. In: Proceedings of the 2nd International Conference on Intelligent Data Engineering and Automated Learning (IDEAL 2000), Data Mining, Financial Engineering, and Intelligent Agents, pp. 386–391. Springer, London (2000). http://dl.acm.org/citation.cfm?id=646287.688630
17. Krötzsch, M., Schaffert, S., Vrandecic, D.: Reasoning in semantic wikis. In: Antoniou, G., Aßmann, U., Baroglio, C., Decker, S., Henze, N., Patranjan, P.L., Tolksdorf, R. (eds.) Reasoning Web. Lecture Notes in Computer Science, vol. 4636, pp. 310–329. Springer, Berlin (2007)
18. Leake, D.B., Kinley, A., Wilson, D.: Acquiring case adaptation knowledge: a hybrid approach. In: AAAI/IAAI, vol. 1, pp. 684–689 (1996)
19. Lieber, J.: Strong, fuzzy and smooth hierarchical classification for case-based problem solving. In: van Harmelen, F. (ed.) Proceedings of the 15th European Conference on Artificial Intelligence (ECAI-02), Lyon, France, pp. 81–85. IOS Press, Amsterdam (2002)
20. Messai, N., Devignes, M.D., Napoli, A., Smaïl-Tabbone, M.: Querying a bioinformatic data sources registry with concept lattices. In: Dau, F., Mugnier, M.L., Stumme, G. (eds.) ICCS, Lecture Notes in Computer Science, vol. 3596, pp. 323–336. Springer, Berlin (2005)
21. Messai, N., Devignes, M.D., Napoli, A., Smaïl-Tabbone, M.: Many-valued concept lattices for conceptual clustering and information retrieval. In: Ghallab, M., Spyropoulos, C.D., Fakotakis, N., Avouris, N.M. (eds.) ECAI, Frontiers in Artificial Intelligence and Applications, vol. 178, pp. 127–131, IOS Press (2008)
22. Minor, M., Bergmann, R., Görg, S., Walter, K.: Adaptation of cooking instructions following the workflow paradigm. In: Marling, C. (ed.) ICCBR 2010 Workshop Proceedings, pp. 199–208 (2010)
23. Pearl, J.: Heuristics—Intelligent Search Strategies for Computer Problem Solving. Addison-Wesley Publishing Co., Reading, MA (1984)
24. Zhang, Q., Hu, R., Namee, B.M., Delany, S.J.: Back to the future: knowledge light case base cookery. In: Schaaf, M. (ed.) Workshop Proceedings of the 9th European Conference on Case-Based Reasoning, pp. 239–248, Trier (2008)
25. Smyth, B., Keane, M.T.: Using adaptation knowledge to retrieve and adapt design cases. Knowledge-Based Systems 9(2), 127–135 (1996)
26. Szathmary, L., Napoli, A.: Coron: a framework for levelwise itemset mining algorithms. In: Ganter, B., Godin, R., Nguifo, E.M. (eds.) Supplementary Proceedings of 3rd International Conference on Formal Concept Analysis (ICFCA'05), Lens, France, pp. 110–113 (2005)
27. Uren, V., Cimiano, P., Iria, J., Handschuh, S., Vargas-Vera, M., Motta, E., Ciravegna, F.: Semantic annotation for knowledge management: requirements and a survey of the state of the art. J. Web Seman. Sci. Serv. Agents World Wide Web 4(1), 14–28 (2006)

Printed in the United States
By Bookmasters